Histology Class Notes

Histology Class Notes

The Big Picture

Tramar F. Murdock, MD

These notes are aimed in organizing and providing general principles of Histology and Histotechnology to my students.

authorHOUSE®

AuthorHouse™ LLC
1663 Liberty Drive
Bloomington, IN 47403
www.authorhouse.com
Phone: 1-800-839-8640

Published by AuthorHouse 08/07/2014

ISBN: 978-1-4969-2793-4 (sc)
ISBN: 978-1-4969-2794-1 (e)

Library of Congress Control Number: 2014912951

CONTENTS

ACKNOWLEDGEMENTS AND SPECIAL THANKS TO:

John Louis Murdock, Sr. BS, HT/HTL (ASCP)
Associates in Tissue Technology, Inc.

Beloved father, friend, mentor

BIBLIOGRAPHY

1. Gartner LP, Hiatt JL, <u>Color Atlas of Histology;</u> 6th edition. Lippincott Williams and Wilkins, Baltimore, MD, 2014.

2. Ross MH, Pawlina W, <u>Histology: A Text and Atlas</u>; 5th edition. Lippincott Williams and Wilkins, Baltimore, MD, 2006.

3. Young B, Heath JW, <u>Wheater's Functional Histology: A Text and Colour Atlas</u>; 4th edition. Churchchill Livingstone, Philadelphia, PA, 2000.

4. Leboffe MJ, <u>A Photographic Atlas of Histology</u>; 2nd edition. Morton Publishing, Englewood, CO, 2013.

HISTOLOGY

INTRODUCTION

HISTOLOGY:

HISTOTECHNOLOGY:

Organelles

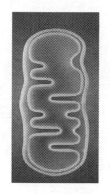

Cells

Human Cardiac Muscle Cells

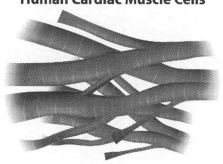

Tissue

Organ

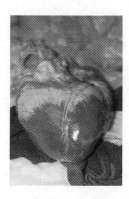

System

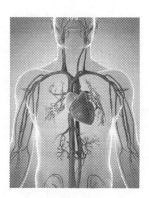

HISTOLOGY

METHODS

(TISSUE PREP/STAINS/MICROSCOPY)

A. **Tissue Preparation**
  1. **Fixation**
      *   The ***FIRST AND MOST IMPORTANT*** step in preparation of a cell, tissue, or organ to preserve structures
      *   Done by heat, freezing, chemicals
      *   ***FORMALIN*** **(AQUEOUS SOLUTION OF FORMALDEHYDE) THE MOST COMMONLY USED FIXATIVE.** Preserves general structure of the cell and extra cellular components
  2. **Processing**
      *   The tissue is washed and dehydrated in a series of alcohol solutions of ascending concentrations to remove water
      *   Organic solvents (xylene, toluene) which are miscible in both alcohol and paraffin (a type of wax) are used to remove the alcohol
      *   Melted paraffin is used to infiltrate the tissue
  3. **Embedding**
      *   The tissue is infiltrated in an embedding medium (ex: paraffin) which is then cooled and hardened in a block
  4. **Cut (Sectioned)**
      *   The cooled block is placed in a special slicing machine (microtome) and cut (steel blades, glass knives)
  5. **Staining**
      *   Sections are colorless, so they must be stained or colored
      *   **HEMATOXYLIN AND EOSIN ARE THE MOST COMMONLY USED STAINS**
      *   **Hematoxylin: Nucleus and nucleolus stain _____**
      *   **Eosin: Cytoplasm, some extra cellular material stain _____**
  6. **Other staining procedures**
      *   To demonstrate certain specific structures (ex: fat, collagen, glycogen) other staining procedures are used

B. **Microscopy**
  A. **The role of a microscope is to magnify an image**
  1. **Various light microscopes**
      *   **Bright-field consist of:**
          o   Light source (sub stage lamp)
          o   Condenser lens (focus beam of light at the level of the specimen)
          o   Stage (slide is placed)
          o   Objective lens (gathers light that passes through the specimen)
          o   Ocular lens (which the image formed by the objective lens may be examined directly)
      *   **Phase contrast-**enables one to examine unstained cells and tissue and living cells
      *   **Dark-field-**no direct light from the light source is gathered by the objective lens
      *   **Fluorescence-**used to display naturally occurring fluorescent (auto fluorescent) molecules to fluoresce under UV light.
      *   **Electron microscope**
      *   **Scanning**

1

HISTOLOGY

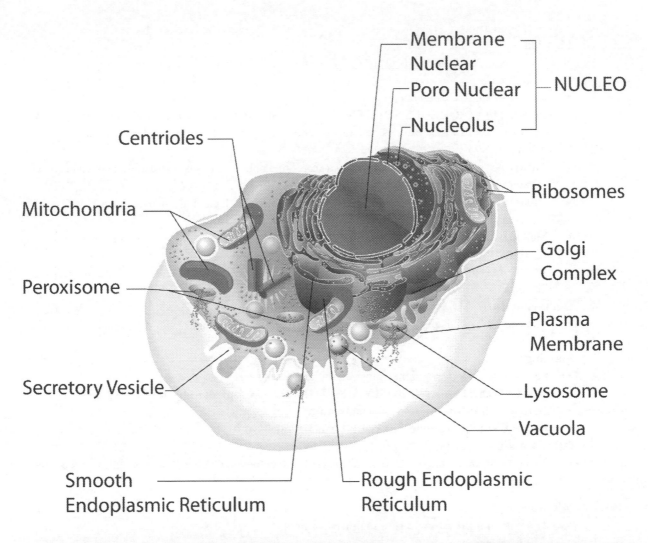

Centrioles

Mitochondria

Peroxisome

Secretory Vesicle

Smooth
Endoplasmic Reticulum

Membrane
Nuclear

Poro Nuclear

Nucleolus

NUCLEO

Ribosomes

Golgi
Complex

Plasma
Membrane

Lysosome

Vacuola

Rough Endoplasmic
Reticulum

Biology   Human Cell

THE CELL

HISTOLOGY

THE CELL

A.   **The Cell**
- **Overview of Cell Structure**
  - o   **See Diagram**

B.   **Plasma membrane (Plasmalemma)**
- **Membrane Transport and Vesicular Transport**
  - o   Simple Diffusion-
  - o   Carrier proteins-
  - o   Channel proteins-
  - o   Endocytosis-
  - o   Exocytosis-
  - o   Pinocytosis-
  - o   Phagocytosis-

C. **Cytoplasm**
- Consist of **material** located **between the plasma membrane and the external environment**

- **Cytosol-**

- **Organelles**
  - o   <u>**Membranous**</u>
    - ▪ **Mitochondria**
      - *   Possess Two Membranes
        1. <u>**Inner (Folds-Cristae)**</u>
           - -   Respiratory Electron-transport chain
           - -   Synthesizing ATP
        2. <u>**Outer (Voltage Channels)**</u>
           - -   Intermembrane space (Enzymes)
           - -   Matrix (Krebs cycle and fatty acid oxidation)

    - ▪ **Endosomes**
    - ▪ **Lysosomes**
    - ▪ **Rough Endoplasmic Reticulum (RER)**
      - *   Cisternae
      - *   Ribosomes
    - ▪ **Smooth Endoplasmic Reticulum (SER)**
      - -   Called ***Sarcoplasmic Reticulum in Muscle***
    - ▪ **Golgi Apparatus**
      - *   cis-Golgi
      - *   trans-Golgi
    - ▪ **Peroxisomes**
      - *   Catalase, peroxidases
      - *   Regulates cellular hydrogen peroxide

C. **Cytoplasm (Cont.)**
- **Organelles**
  - o <u>**Nonmembranous**</u>
    - ▪ **Cytoskeleton**- contain **filamentous proteins**
      - * **Microtubules**
        - - Hollow tubular structures (**α, β tubulin**)
        - - Maintain cell shape and movement of organelles and inclusions
        - - Spindle fibers of mitosis
        - - **In cilia and flagella**
      - * **Filaments**
        - - **Thin (*Actin) filaments***
          - Anchor and movement
          - Structural core of microvilli
        - - ***Intermediate Filaments***
          - Shape, Support
        - - ***Thick (Myosin) filaments***
          - Facilitate cell movement
    - ▪ **Centrioles**
      - * Made of **9 triplets of microtubules**
      - * Migrate to opposite poles of cell during cell division **(Mitotic spindle formation)**
      - * Provide **Basal bodies** (cilia and flagella) (**9 + 2 arrangement**)

    - ▪ **Centrosomes**

    - ▪ **Ribosomes**
      - * Essential for **protein synthesis**

    - ▪ **Inclusions**

D. **Nucleus**
- **Chromatin-**
  - o **Euchromatin**
  - o **Heterochromatin**
- **Nucleolus-**
- **Nuclear envelope**
  - o **Two membranes** with perinuclear cisternal space which are continuous with the cisternal space of the rER.
  - o **Nuclear pores**
  - o Disassembles during cell division (chromosome separation)
- **Nucleoplasm:**

E. **Cell Cycle and Mitosis**
- **Interphase**
  - o **G1 phase:**
  - o **S phase:**
  - o **G2 phase:**
  - o **G0 phase:**
- **Mitosis: prophase, prometaphase, metaphase, anaphase, telophase, cytokinesis**

HISTOLOGY

TISSUES: Concepts and Classification

Tissues are groups or aggregates of cells which are organized to perform one or more specific functions.

ALL ORGANS ARE MADE UP OF FOUR BASIC TISSUE TYPES !!!!

1. *Epithelium (epithelial) Tissue*
   - Covers the body surfaces, lines body cavities, forms glands
   - Identified by closely packed, contiguous cells and the presence of a free surface
   - Usually joined by specialized cell-to-cell junctions
   - Subclassifications based on shape (squamous-flat), cell layers (stratified-multiple layers)

2. *Connective Tissue*
   - Underlies and/or supports
   - Identified by its extracellular matrix
   - Unlike epithelial cells, CT cells are conspicuously separated from one another
   - The spaces between the cells are filled with material produced by the cells

3. *Muscle Tissue*
   - Made up of contractile cells and is responsible for movement
   - Categorized on the basis of a functional property (the ability of its cells to contract)
   - Cells characterized by large amounts of actin and myosin
   - Smooth, skeletal, cardiac

4. *Nerve tissue*
   - Receives, transmits, and integrates information from outside and inside the body to control activities of the body
   - Consists of nerve cells (neuron) and supporting cells (neuroglia, Schwann)

HISTOLOGY

EPITHELIUM

A. **Epithelium**
   - **Lines and covers virtually all free surfaces of the body**
   - <u>Three Principle Characteristics</u>
     o **Have cell junctions**
     o **Polarity (depending on surface domains)**
     o **Basement membrane**
   - Creates a barrier
   - **Avascular!**

B. **Classification**
   - ***Based on Cell Layers and Cell Shape***
     o **Simple-One Layer Thick**
       ▪ **Simple Squamous** (endothelium, alveolar of lung, mesothelium)
       ▪ **Simple Cuboidal** (ducts, thyroid, kidney)
       ▪ **Simple Columnar** (intestine, colon, gallbladder)
     o **Stratified- Two or more Layers Thick**
       ▪ **Stratified Squamous Non keratinizing** (oral cavity, esophagus, vagina)
       ▪ **Stratified Squamous Keratinizing** (skin)
       ▪ **Stratified Cuboidal** (ducts of salivary glands, sweat glands)
       ▪ **Stratified Columnar** (portions of male urethra, anorectal junction)
     o **Other Epithelia**
       ▪ **Transitional** (bladder-nondistended)
       ▪ **Pseudostratified Ciliated Columnar** (respiratory, epididymis)

C. **Functions**
   - **Secretion**
   - **Absorption**
   - **Transport**
   - **Protection**
   - **Receptor function**
   - **Form glands**

D. **Cell Polarity**
   - **Apical (free) Surface specializations/Domain**
     o **Microvilli**: fingerlike projections of cell membrane; absorption and secretion
     o **Stereocilia**: elongated microvilli; facilitate absorption
     o **Cilia**: contain microtubules; movement

D. Cell Polarity (Cont.)
- **Lateral Surface/Domain**
    - o **Occluding Junctions**
        - **Zonula occludes (ZO)** : tight junctions; prevents leakage

    - o **Anchoring Junctions**
        - **Zonula adherens (ZA):** beneath ZO.; serves in attachment of adjacent cells
        - **Macula adherens (or desmosomes):** 2 disc shaped plaques; spot like adhesion

    - o **Communicating Junctions**
        - **Gap junctions (or nexus):** sites of intercellular communication

- **Basal Surface/Domain**
    - o **Basement membrane/Basal lamina: attaches epithelium to underlying connective tissue**
    - o **Provides support, filtration, cell-to-cell interactions**
        - **Lamina lucida:**
        - **Lamina densa:**
        - **Lamina reticularis:**

    - o **Anchoring Junctions**
        - **Hemidesmosomes**
        - **Focal Adhesions**

    - o **Lamina propria**
        - Consist of collagen, elastic fibers, matrix material, fibroblasts, blood, and lymphatic vessels

HISTOLOGY

EPITHELIUM

E. **Glands**
- **Epithelial down growths**
- **Specialized in producing secreted material**

    1. Exocrine Glands
- o Regulated by response to nerve impulses and certain hormones
- o Consist of **a gland (secretory unit) and a *DUCT**
- o **Classification based on**:
  - **Duct arrangement**
    - * Simple (sweat gland)
    - * Compound (pancreas)
  - **Shape of secretory unit**
    - * Tubular
    - * Acinar
    - * Tubuloalveolar
  - **Secretion type**
    - * Serous
    - * Mucous
    - * Mixed
  - **Secretion Production**
    - * Merocrine (pancreas)
    - * Holocrine (sebaceous glands)
    - * Apocrine (mammary glands)

    2. Compound Exocrine Glands
- o **Gland (secretory unit and ducts) + stroma** (supporting connective tissue)
- o Includes **blood vessels, nerve fibers** (submandibular gland)
  - Interlobular ducts
  - Intralobular ducts
  - Interlobular septum

    3. Endocrine Glands
- o **Similar organization but NO DUCTS!!!!!**
- o **Release secretions into blood stream**
- o Secretions known as **hormones**
- o Store secretions they synthesize; **discharge intermittently by exocytosis** (thyroid gland)

    4. Myoepithelial Cells
- o **Contractile** epithelial cells (sweat, salivary, mammary glands)

HISTOLOGY

CONNECTIVE TISSUE

A. **Connective Tissue**
- Role is to **support, interconnect, nourish other tissues; storage; protect**
- Consists of an <u>**extracellular matrix**</u> (fibers, ground substance, tissue fluid) <u>**and cells**</u> (fibroblasts, macrophages, mast cells, etc.,)

B. **Embryonic Connective Tissues**
- **Mesenchyme: embryonic tissue;** differentiate into cells typical of adult connective tissue; in adults **mesenchyme** role important in tissue repair with fibroblasts
- **Mucous: deep to feta skin and umbilical cord; Wharton's jelly**

C. **Adult Connective Tissues**
- **Extracellular Matrix (ECM)** consist of <u>**Three Types of Fibers**</u>
  o <u>**Collagen Fibers (Type I)**</u>
    ▪ **MOST ABUNDANT**! wide, wavy, pink
    ▪ Inelastic, **tensile**
    ▪ Made of many collagen molecules packed together
  o <u>**Reticular Fibers (Type III)**</u>
    ▪ **Finely branched**, tightly wove
    ▪ Consist of collagen fibrils
    ▪ **Mesh-like pattern**
    ▪ Network for **hematopoietic and lymphatic tissue**
  o <u>**Elastic Fibers**</u>
    ▪ **Straight and narrow**, composed of **elastin**
    ▪ Thinner than collagen fibers
    ▪ Branching pattern
    ▪ **Stretch, distention, resilience of tissue**
    ▪ Ligaments, larynx, elastic arteries
- **Ground Substance (Amorphous Component)**
  o Space between cells and fibers
  o Lost in H&E staining
  o Rich in **proteoglycans and hyaluronic acid**
  o Fluid content **facilitates diffusion** of nutrients and waste products
- **Connective Tissue Cell Types**
  o **Fibroblast**-synthesize collagen, elastin, reticular fibers; elongated structure
  o **Macrophages (histiocytes)**-phagocytic cells
  o **Adipocytes**-store fats
  o **Mast cells**-vasoactive and immunoreactive substances
  o **Undifferentiated mesenchymal cells**- repair and formation of new tissue (wound healing and neovascularization)
  o **Lymphocytes**-immune response; T, B, and NK cells
  o **Plasma cells**-antibody producing cells from B lymphocytes
  o **Neutrophils**- acute inflammation

- **Connective Tissue Cell Types (Cont.)**
  - o **Eosinophils**-allergic reactions
  - o **Basophils**- allergic reactions
  - o **Monocytes** (from bone marrow)

D. **Adult Connective Tissue Types**
  - **Loose (Areolar) Connective Tissue**
    - o Distributed widely
    - o Connective tissue component of serous and mucous membranes
    - o Makes up much of the superficial fascia; wraps neurovascular bundles
  - **Dense Connective Tissue**
    - o <u>Regular</u>-thick, orderly parallel fibers; tendon, ligaments; densely packed, little ground substance
    - o <u>Irregula</u>r-haphazardly arranged fibers; dermis of the skin

E. **Special**
  - Adipose
  - Blood
  - Cartilage
  - Bone

HISTOLOGY

ADIPOSE TISSUE

A. **Adipose Tissue**
- **A specialized type of connective tissue made up of fat storing cells call adipocytes**
  - **Two Types of Adipose Tissue White**
    - **-Primarily in Adults**
      - * Store energy, insulates, secretes hormones
      - * **Unilocular**
      - * Produces **hormone LEPTIN***
      - * Derived from undifferentiated mesenchymal cells; lipoblasts→adipocytes
      - * **Spherical shape, nucleus pushed to side**
      - * Forms hypodermis in connective tissue

    - **Brown-Fetal Life**
      - * Diminishes in amount
      - * Multilocular
      - * Nucleus central, empty vacuoles in cytoplasm
      - * Generates heat (Nonshivering thermogenesis)

B. **Leptin**
- **Inhibits appetite; an appetite suppressant**
  - Produces **satiety**
  - If absent, increased appetite →increased obesity

HISTOLOGY

CARTILAGE TISSUE

A. **Cartilage**
- **Special Type of Connective Tissue**
- Develops from **mesenchymal cells→chondroblasts→chondrocytes** (in lacunae)
- Chondroblasts secrete **specialized extracellular matrix** (more **glycoprotein +connective tissue; elastin**)
- Chondrocytes grows and divide→**isogenous** group of cells
- **AVASCULAR!!!!!!**

B. **Three Types of Cartilage**
- **Hyaline Cartilage**
  - **Most abundant**
  - **Type II collagen fibers, proteoglycans, hyaluronic acid**
  - **Glassy appearance** (gross)
  - Chondrocyte within a lacunae
  - **Hydrated** (permits diffusion of small metabolites; increases resiliency)
  - Main function is **provide sliding area for joints**, cushions
  - Location: nose through bronchi, articular surfaces, synovial joints
- **Elastic Cartilage**
  - **Type II collagen fibers +Elastin** in matrix
  - **More pliable**
  - Location: external ear, auditory canal
- **Fibrocartilage**
  - **Dense regular connective tissue and Hyaline cartilage (Types I and II)**
  - **Chondrocytes "mixed"** among collagen fibers (single, rows, groups)
  - Act like **"shock absorbers"**
  - Location: intervertebral disc, pubic symphysis

C. **Ability for Cartilage to Repair**
- **LIMITED!!!!!!**
- Due to **AVASCULARITY** AND **IMMOBILITY** OF THE CHONDROCYTES

D. **Growth**
- **Interstitial (within)** and **appositional (peripheral)**

HISTOLOGY

BONE

A. **Bone**
- **Provides support and protection;** stores **calcium and phosphate**
- Contains living cells **(osteocytes), Type I collagen**, both in an extracellular matrix
- **Matrix heavily calcified (HYDROXYAPATITE crystals)**
- **VERY VASCULAR!!!!!!! (Lots of capillaries)**

B. **Cell Types**
- **Osteoprogenitor cell**-mesenchymal origin ;resting cell ;**transforms into osteoblast**
- **Osteoblast- "immature"** bone forming cell that **secretes unmineralized bone matrix (osteoid)**
- **Osteocyte- "mature"** bone forming cell surrounded **by MINERALIZED** bone matrix
- **Osteoclast-phagocytic cell, bone resorbing cell; Howship's Lacunae**

C. **Osteocyte Survival**
- **Haversian System**-lamella (concentric ring of bone tissue, lacunae with osteocytes found between lamella)
- **Haversian canal** (surrounded by lamella; house blood vessels, nerve fibers, cells)
- **Volkmans canals** (transverse channels in lamella bone which communicate with Haversian canals)
- **Cananiculi** (tiny channels) connect lacunae with one another and link to bone surfaces bathed by tissue fluid. Cananiculi bring tissue fluid, oxygen, nutrients to osteocytes

D. **Bone Membranes**
- Two fibrous membranes
  - **Periosteum: surrounds bone;** fibrous layer and cellular layer; **Sharpey's fibers**
  - **Endosteum: lines inside surfaces**

E. **Bone Tissue Classification**
- **Grossly**
  - **Compact (Dense)**-area with no cavities
  - **Spongy (Cancellous)** -interlacing cavities
- **Histologically**
  - **Immature**-No ORGANIZED lamellate arrangement
    - More cells
    - Cells randomly arranged
    - Not heavily mineralized
  - **Mature**-Made up of osteons (Haversian System-cylindrical units)
- **Shape**
  - **Long bones**-longer, consist of shaft **(diaphysis)** and two ends **(epiphysis)**; Tibia
  - **Short bones**-equal in length ; Bones of the hand
  - **Flat bones**-thin, plate like; Skull
  - **Irregular bones**-Vertebra, Ethmoid

F. **Bone Formation**
- **Two Mechanisms**
  - o **Intramembranous Ossification-mesenchymal cells <u>transform</u> into osteoblast.** (Flat bones)

    - **Primary Ossification Center-**

  - o **Endochondral Ossification-<u>existing cartilage replaced by bone forming cells</u>** called osteoblasts. (Long bones)

    - **Primary Ossification Center-**

    - **Secondary (Epiphyseal) Ossification Center-**

    - **Epiphyseal Plate-**Responsible for **lengthening of long bone**
      * **Zone of Reserve Cartilage:**
      * **Zone of Proliferation:**
      * **Zone of Hypertrophy:**
      * **Zone of Calcification:**
      * **Zone of Ossification:**
      * **Zone of Erosion:**

HISTOLOGY

BLOOD

A. **Blood**
- "Fluid" Connective Tissue; circulates through the cardiovascular system

B. **Function**
- **Transports nutrients and oxygen to cells**
- **Transports waste and carbon dioxide away from cells**
- Delivers **hormones**
- Acts as a **buffer**
- Helps maintain **homeostasis**

C. **Consists of Cells**
- **Erythrocytes** (RBC's)-most abundant
- **Leucocytes** (WBC's)
- **Platelets**

D. **Plasma**
- The **Clear liquid part**; **rich in proteins** (albumin, globulins, fibrinogen),**water, electrolytes, oxygen, carbon dioxide, nitrogen**

E. **Serum**
- **Straw colored; Absent fibrin (fibrinogen) or components necessary for clotting**

F. **Cell Types**
- **Erythrocytes (RBC)**
  - o **Anuclear**; biconcave disc; 7-8um in diameter
  - o Binds oxygen for delivery to the tissues
  - o Binds carbon dioxide for removal
  - o Contains **Hemoglobin (Hgb)**
    - ▪ Protein molecule which consists of 4 subunits, each containing a heme group   containing iron.
    - ▪ Protein to transport oxygen and carbon dioxide
    - ▪ **HbA, HbA2,HbF**
- **Reticulocytes**
  - o IMMATURE RBC's
- **Platelets**
  - o Also called **thrombocytes**
  - o Produced by **megakaryocytes**
  - o **Clot formation**

F. **Cell Types (Cont.)**
- **Leukocytes: Two Groups (based on presence of granules)**
  - o **Granulocytes**
    - **Neutrophils**
      - * **MOST NUMEROUS WBC**
      - * **Multilobed** (Polymorphonuclear neutrophils)Barr body (in females)
      - * **Primary**/Azurophilic granules(Myeloperoxidase)
      - * **Secondary**/Specific granules (Lysosomes)
      - * **Tertiary granules** (Phosphatases, gelatinases)
      - * **Motile cells** (via chemotaxis), active phagocytes
    - **Eosinophils**
      - * Large **RED GRANULES** in cytoplasm
      - * **Bilobed nucleus**
      - * **Primary**/Azurophilic granules
      - * **Secondary**/Specific granules
      - * Associated with **allergic reactions, parasitic infections chronic inflammation, neoplasms**
    - **Basophils**
      - * **Least numerous** of the WBC's
      - * Granules stain **BLUE**
      - * **Primary**/Azurophilic granules (lysosomes)
      - * **Secondary**/Specific granules (vasoactive agents)
      - * Role **similar to mast cells**

  - o **Agranulocytes: Unlobed nuclei with Primary/Azurophilic granules**
    - **Lymphocytes**
      - * **MOST COMMOM** agranulocyte
      - * Cells that are able to **recognize and respond to antigens**
      - * **Three Main Types**
        - **T cells** :long life,**(CD4,CD8)**, cell-mediated immunity
          - o **CD4 (Helper)**
          - o **CD8 (Cytotoxic)**
        - **B cells**: short life, produce circulating antibodies**; plasma cells**
        - **NK cells**: programmed to kill certain virus-infected cells and certain **tumor cells**
    - **Monocytes**
      - * **Largest of the WBC's in blood smear**
      - * Bone marrow→Blood→Body tissue→ differentiates→ **(histiocytes,osteoclasts, macrophages,)**
      - * **Phagocytic cells**

    - **Platelets (Thrombocytes)**
      - * **Membrane bounded, anucleate, cytoplasmic fragments**
      - * Derived **from megakaryocytes**
      - * Important in **blood clot formation**

G. **Bone Marrow Composition**
  - In marrow cavity of long bones
  - **Very vascular**
  - <u>**Two forms**</u>
    - o **Red marrow**
    - o **Yellow marrow**

H. **Formation of Blood Cells (Hemopoiesis)**
  - Includes :red blood cells (RBC's), white blood cells (WBC's), platelets
  - **Adult**
    - o Erythrocytes, granulocytes, monocytes, platelets all formed in the Red Bone Marrow
    - o Monocytes also formed in the red bone marrow and lymphatic tissues

  - **Fetal development**: RBC's and WBC's formed in different organs
    - o Yolk sac phase -FIRST!!!!! (Blood islands)
    - o Hepatic phase-SECOND!!!
    - o Bone marrow phase

I. <u>**BLOOD CELLS ARE DERIVED FROM A COMMON STEM CELL: PLURIPOTENTIAL CELL**</u>

  - **Myeloid stem cells and Lymphoid stem cells→→→→Progenitor cells**

HISTOLOGY

MUSCLE TISSUE

A. **Muscle**
- Responsible for **contraction, movement**

B. **Structural/Functional Subunit of the Muscle Fiber (aka Muscle Cell) is the Myofibril**
- **Myofilaments**: individual polymers of **thick (myosin) and thin (actin)** filaments
- **Myofibrils**: Bundles of myofilaments, **"Contractile elements"**
- **Muscle fibers (cell)** :made up of bundles of **myofibrils**
- **Muscle fascicles** :Bundles **of muscles fibers**

C. **Three Main Types of Muscle**
- **Skeletal Muscle**
  - **Voluntary control**
  - **Multinucleated, nuclei on "edge"**, beneath the plasma membrane
  - **Elongated** fibers
  - **Striations** (alternating dark and light transverse bands)
  - **Dark bands-(A Bands)**
  - **Light bands-(I bands)**
  - Consist of fibers held together by connective tissue
    - **Endomysium**
    - **Perimysium**
    - **Epimysium**
  - **Contractile unit** of the skeletal muscle is **the SARCOMERE**
    - Sarcomere is a segment of the myofibril **between two adjacent Z lines**
    - Sarcomere made **of thin filaments (actin) and thick filaments (myosin)**
    - **Regulation of contraction involves:**
      * **Calcium**
      * **Sarcoplasmic reticulum (SR)**
      * **Transverse(T)-tubules**
    - **Contraction Cycle (Sliding Filament Theory)**
      * **Thin filaments** move along **Thick filaments** in **5 Stages**
        1. **Attachment**: myosin head binds tightly to actin (ATP absent)
        2. **Release**: myosin head uncoupled from thin filament (ATP binds to myosin)
        3. **Bending**: myosin head moves
        4. **Force generation**: myosin head releases phosphate
        5. **Reattachment**: myosin head binds to new actin molecule
  - **Motor innervation** via **neuromuscular junction** which is the contact between terminal branches of the axon and muscle
  - Limb muscles, back, abdomen, digits

- **Cardiac Muscle**
  - o **Uninucleate, in center** of cell
  - o **Sarcomere arrangement**
  - o **Intercalated disc** (attachment site between cardiac muscle cells)
  - o **Involuntary control, rhythmic contraction**
  - o Heart **(myocardium)**

- **Smooth Muscle**
  - o **Elongated cells with tapered ends**
  - o **Nuclei in center** of cell
  - o Contractile proteins **(Dense bodies and filaments)**
  - o **No striations**
  - o **Involuntary control**
  - o Viscera, uterus, intrinsic eye muscles

D. **Special Receptors** -vessels, organs, arrector pili muscle
  - **Muscle Spindles**
    - o **Encapsulated: Proprioception;** respond to **length and rate of change** in muscle
  - **Golgi tendon**
    - o **Encapsulated: Proprioception;** respond to changes in **tension** and **rate of tension change** around a **joint**

HISTOLOGY

NERVOUS TISSUE

A. **Nerve Tissue**
- Enables the body to respond to continuous changes in the external and internal environment
- Divided
  - o **Anatomically**
    - **Central nervous system (CNS)**: brain, spinal cord
    - **Peripheral nervous system (PNS)**: cranial nerves, spinal nerves, ganglia.
  - o **Functionally**
    - **Somatic nervous system (SNS)**
      - * Consisting of body parts of the CNS and PNS
      - * Sensory and motor innervation to all parts of the body, except the viscera, smooth muscle, and glands
    - **Autonomic nervous system (ANS)**
      - * Involuntary motor innervation to smooth muscle, conducting system of the heart, and glands
      - * Further divided Sympathetic Division and Parasympathetic Division

B. **Nervous Tissue Elements**
- **The Neuron**
  - o **Functional unit** of the Nervous System. **DO NOT REPLICATE!**
    - **Cell body (Perikaryon)** contains the nucleus, nucleolus, and other organelles.
    - **Nissl bodies**-stacks of rER.
    - **Axon hillock**-absence of Nissl bodies; void of impulses
  - o **Dendrites**
    - Receptors that **receive stimuli** from other neurons or the external environment; conduct **impulses toward the cell body**
  - o **Axon**
    - **Longest process. ONE AXON/NEURON!**
    - **Transmits stimuli to other neurons or effector cells (ex. Muscle)**
  - o **Synapses (Synaptic junctions)**
    - Where **neurons communicate** with other neurons and effector cells
    - **Specialized junction**
  - o **Structural Types of Neurons**
    - **Unipolar:**
    - **Bipolar:**
    - **Multipolar:**
    - **Pseudounipolar:**
- **Supporting Cells**
  - o **Nonconducting cells**. Provide protection, insulation, metabolic exchange pathways
  - o **Peripheral Nervous System (PNS)**
    - **Schwann Cells**
      - * Form **myelin sheaths** around **axons**.(Increases the conduction velocity of the impulse along the axon.)
      - * **Neurilemma**
      - * **Node of Ranvier**: the region where the myelin sheath of one Schwann cell ends and the next one begins

- **Supporting Cells (Cont.)**
  - o **Peripheral Nervous System (PNS)**
    - **Satellite Cell**
      - * **Within ganglia**, surround the nerve cell bodies (contains the nucleus)
      - * Provides insulation, pathway of metabolic exchange
  - o **Central Nervous System (CNS) (Neuroglia or glia cells)**
    - **Oligodendrocytes**: form and maintain **myelin; "fried egg"** appearance
    - **Astrocytes: "star shaped"**; physical and metabolic support for the neurons
    - **Microglia cells**: inconspicuous cells; which have **phagocytic properties**
    - **Ependymal cell**: columnar in shape with **cilia and microvilli**; fluid-transporting cells; **absorb the cerebrospinal fluid (CSF)**

C. **Organization of Spinal Cord and Brain**
  - **Spinal Cord**
    - o **Flattened cylindrical structure**; "butterfly" appearance
      - **Gray matter-centrally located; cell bodies and their dendrites**
      - **White matter-peripherally located**; contain **myelinated and unmyelinated axon**s
  - **Medulla Oblongata**
    - o Inferior part of brain; joins spinal cord
      - **White matter-Contains** ascending and descending **tracts and several nuclei**
  - **Cerebrum**
    - o **Cortex-Gray matter** on surface; divided into **six layers**
      - **Molecular layer: most superficial**; neuronal axons and dendrites; glial cells
      - **Outer granular layer: pyramidal cells** and **stellate cells**
      - **Pyramidal cell layer: pyramidal cells**
      - **Inner granular layer: stellate cells**
      - Ganglionic layer: **Betz cells,** stellate cells, **cells of Martinotti**
      - Multiform layer: **fusiform cells**
    - o **White matter**-deeper, mostly **myelinated fibers** and associated neuroglial cells
  - **Cerebellum**
    - o Coordinates **voluntary movement and muscle function** to maintain balance, normal posture
      - **Cortex: Gray matter**
        - * **Molecular layer**: outer; unmyelinated fibers; few neurons
        - * **Granular layer**: inner; many neurons
        - * **Purkinje cells**
      - **Medullary Substance**
        - * Internal **white mass**
  - **Meninges**
    - o **Brain and the spinal cord** are **wrapped in connective tissue** and s**urrounded by fluid (CSF)**
    - o Layers**: Dura mater, Arachnoid mater and Pia mater**
    - o **Blood- brain barrier**: formed by **tight junctions** between capillary endothelial cells and astrocytes
      Restricts passage of substances i.e.drugs, from the blood stream to the brain

D. **Organization of the Peripheral Nervous System (PNS)**
  - **Peripheral Nerves**
    - o Composed of **numerous nerve fibers collected in fascicles (bundles)**
    - o Connective tissue of **the PNS**
      - **Endoneurium:** the connective tissue surrounding each fiber
      - **Perineurium:** the connective tissue which surrounds a fascicle.(bundle)
      - **Epineurium:** the connective tissue which surrounds an entire peripheral nerve

HISTOLOGY

LYMPHATIC SYSTEM

- **Lymphatic system consists of groups of cell, tissues, and organs that monitor the body surfaces and internal fluid compartments and react to the presence of potentially harmful substance**
- **Lymph-** a slightly **yellow liquid** found in the lymphatic vessels and derived from the tissue fluids. It consists of a liquid portion and of cells, most of which are lymphocytes. It is collected from all parts of the body and returned to the blood via the lymphatic system

A. **Lymphoid Tissue and Immune System**
- **Lymph Nodes**
    - o **Filter lymph (fluid)** passing through lymphatic's before its return to the blood
    - o Part of the immune system
- **Lymphoid Organs**
    - o **Thymus**: **T cell production**
    - o **Lymphoid follicles(or Lymph Nodules): unencapsulated group of mostly B cells**
    - o **Lymph nodes: encapsulated**, bean shaped structures that filter the lymph
    - o **Spleen**: largest lymphatic organ
- **Lymphatic Tissue**
    - o Scattered in various organs; heavily **populated with lymphocytes**

B. **Immune Responses**
- **Humoral**
    - o **B cells→Plasma cells →antigen-specific immunoglobulins (antibodies)**
    - o **B cells →**memory cells
- **Cell mediated**
    - o **T cells involved in cell-mediated responses; do not synthesize Ig**
    - o **T cells→Helper T (TH) cells**
    - o **T cells→Cytotoxic T (TC) cells**
    - o **T cells→ Memory T (TM) cells**

C. **Lymphocytes**
- **B lymphocytes (Bursa)**
    - o Surface membrane Ig specific for that antigen
    - o Many mature into plasma cells
    - o Memory B cells
- **T lymphocytes**
    - o Do not synthesize Ig
    - o T cell antigen receptor on cell surface different from B cell
    - o **CD4, CD8**
    - o **Cytotoxic T cells (Killer cells)**
- **Natural Killer (NK) Cells**
    - o **Lack antigenic markers** like B and T cells
    - o Kill cells coated with antibodies

D. **Supporting cells (Antigen-presenting cells/APCs)**
   o   Macrophages and dendritic cells

E. **Lymphoid Organs**
   •   **Thymus**
      o   **Active in children**
      o   **Bilobed**
      o   **T Cells** (thymocytes)
      o   **Epithelioreticular cells**: form blood-thymus barrier (Reticular network)
      o   Supported by CT **capsule and septa**
         ▪   **Cortex**
            *   **Outer region**; predominantly **developing thymocytes**
            *   Darker staining
         ▪   **Medulla**
            *   **Inner region**; fewer **thymocytes**
            *   Lighter staining
            *   <u>**HASSALL'S CORPUSCLES**</u>-Keratinized epithelial cells; concentrically arranged; Produce chemicals **(Interleukins)** for T cell maturation
      o   Incompletely subdivided into lobes by septa

F. **Lymphoid Follicles (Lymphatic nodules)**
   •   **Scattered** in various organs
   •   **No capsules**
   •   **GALT:_____ MALT:_____BALT:_____**
   •   Most small and discrete :urinary tract, respiratory system
   •   Some large: tonsils, Peyer's Patches, ileum, appendix
   •   Lighter center staining-germinal center: activated B cells

G. **Lymph Nodes**
   •   **Filters consisting of B cells, T cells, macrophages**
   •   **Bean shaped**
   •   Dense connective tissue **capsule**
      o   Divides into **trabeculae**
   •   Supported by **reticular** connective tissue
   •   **Cortex**
      o   **Lymphoid follicles** with **germinal centers**
         ▪   <u>**Primary Follicles**</u>**: uniform; small inactive** B cells
         ▪   <u>**Secondary Follicles**</u>
            *   **Mantle zone: darker zone; inactive** B cells
            *   **Marginal zone**: lighter zone around mantle
   •   **Paracortex**
      o   **T cells dominate**
      o   **High endothelial venules (HEV) located**

   •   **Medulla**
      o   **Channels(cords)** packed **with lymphocytes**
      o   T cells
      o   Plasma cells from cortex secrete **antibodies into lymph**
      o   **Afferent lymph vessels and efferent lymph vessels**

H. Spleen
- **Largest** lymphoid organ
- Responds to **blood antigens**; **phagocytosis** of old RBC's
- **Capsule**
  - o **Fibrous CT and smooth muscle(myofibroblasts)**
  - o **Trabeculae** penetrate spleen from capsule
  - o **Reticular fibers** form framework
- **Splenic Pulp**
  - o <u>**NOT DIVIDED INTO CORTEX OR MEDULLA OR SUBDIVIDED INTO LOBULES**</u>
    - ▪ <u>**Red Pulp**</u>
      - * **Major portion; Erythrocytes**
      - * **Venous sinuses and splenic cords**
      - * Plasma cells
    - ▪ <u>**White Pulp**</u>
      - * **Pale area**
      - * **Lymphocytes aggregate**
      - * **T cells** form **periarterial lymphatic sheaths (PALS)** around **central arteries**
      - * **Splenic nodules**: B cells in germinal centers
      - * Macrophages

HISTOLOGY

CARDIOVASCULAR SYSTEM

A. **The Cardiovascular System**
- **Transport system** that **carries blood and lymph to and from the tissues of the body**
- **Components**: **Heart, blood vessels,** and **lymphatic vessels**

B. **Basic Structure of Blood Vessels**
- **Histology**
  - **Walls Composed of Three Layers:**
    - **Tunica (interna) intima**
      - * **Inner most layer** (luminal side)
      - * Single layer of **simple squamous epithelium** call **endothelium**
      - * Basal lamina
      - * Subendothelial layer
    - **Tunica media**
      - * **Middle layer**
      - * **Smooth muscle**
      - * Bordered by internal and external **elastic membrane**
    - **Tunica (externa) adventitious**
      - * **Outer layer**
      - * Primarily of longitudinal **collagenous tissue** and a **few elastic fibers**
      - * May contain **blood vessels and nerves**

C. **Types of Arteries**
- **Arteries** transport blood **away from heart** at **high pressures**
- **Tunica media thickest** layer
  - **Elastic arteries**
    - **Largest** of all arteries
    - Aorta and its branches, Pulmonary arteries
    - Serve primarily **as conduction tubes**
    - **Tunica media**: contain **multiple layers of elastic lamellae** separated by **smooth muscle**
    - **Tunica adventitia is relatively thin;** often has **vasa vasorum**
  - Muscular arteries
    - **Tunica media** up to **40 layers of smooth muscle ; less elastin**
    - **Tunica intima thinner**; contains a more prominent internal elastic membrane
    - **Tunica adventitia relatively thick**; often separated from the media by a recognizable external elastic membrane
  - Small arteries and arterioles
    - **Small artery** may have **up to 8 layers of smooth muscle** in the **tunica media**
      The **source of vascular resistance; Hypertensive changes**
    - **Arterioles** have **one to 2 layers of smooth muscle** in the **tunica media**
    - **Metarterioles** serve as flow regulators to the capillary beds

D. **Capillaries**
- **Smallest diameter** blood vessels
- Consists only of **Endothelium and basal lamina**
- Often smaller than the diameter of a red blood cell
- **Site of diffusion** of oxygen, CO2 and other nutrients
  - o **Three Types**
    - **Continuous: Continuous** with basal lamina; CNS, muscle, CT, lung
    - **Fenestrated**: **Perforations (pores):** endocrine glands, kidneys, (GI tract)
    - **Discontinuous (sinusoidal):** with basal lamina: liver, spleen, bone marrow

E. **Veins**
- Carry blood **toward the heart under low pressure**
- **Tunica of veins are not as distinct** or well defined as arteries; Walls **thinner**
- Many supplied with **valves**
  - o **Types of Veins**
    - **Venules**
      - * **Smallest;** structurally similar to capillaries
      - * **Pericytes** present but **replaced by smooth muscle in tunica media**
      - * High endothelial venules in certain lymphatic organs
    - **Medium veins**
      - * **Less than 1 cm** in diameter
      - * **Tunica interna** consist of **endothelium and CT**; no internal elastic membrane
      - * **Tunica media** consists of **smooth muscle**
      - * **Tunica externa (adventitia):** collagen, elastic fibers, smooth muscle
    - **Large veins**
      - * **Thick tunica interna**
      - * **Thin tunica media**
      - * **Tunica externa thickest layer**
      - * Superior Vena Cava, Inferior Vena Cava.

F. **The Heart**
- **Wall** composed of **Three Layers**
  - o **Endocardium**
    - **Lining of atria and ventricles**
    - **Simple squamous endothelium** as well as a **subendothelial fibroelastic CT**
    - Contraction synchronized by **specialized muscle fibers**
      - * **SA node**
      - * **AV node**
      - * **Bundle of His**
      - * **Right and Left bundle branches**
      - * **Punkinje fibers**
  - o **Myocardium**
    - Consists of **cardiac muscle**
  - o **Epicardium (Visceral Pericardium)**
    - **Outer surface** of the heart
    - **Simple squamous mesothelium**
    - Connective tissue, nerves, and blood vessels that supply the heart
- **Heart Valves**
  - o **Connective tissue** with an **overlying endocardium**

HISTOLOGY

INTEGUMENTARY SYSTEM

A. **Integument**
- **Skin and its derivatives** (hair, nails, sebaceous glands, arrector pili muscle, sweat glands)
- **Protective barrier, absorption, excretion, sensation, thermoregulation, Vitamin D synthesis**

B. **Skin**
- **Categorized as Thin or Thick**
  - o **Epidermis-stratified squamous <u>keratinized</u> epithelium**
  - o **Dermis-dense irregular collagenous c**onnective tissue
  - o **Hypodermis (Superficial fascia)/subcutaneous fat**
- **Epidermis and dermis interdigitate** to form **dermal papillae**

C. **Layers of the Epidermis**
- **Stratum basale (Stratum germinativum)**
  - o Rest on **basement membrane**
  - o <u>**Mitotically active**</u> **cuboidal to columnar** cells; **stem cells**
  - o **Keratinocytes (<u>most numerous</u>)**
- **Stratum spinosum (Prickly layer)**
  - o **Short processes (intercellular bridges)**; attach to other cells by **desmosomes**
- **Stratum granulosum**
  - o Accumulated <u>**keratohyalin granules;**</u> thin in **Thin skin**
- **Stratum corneum**
  - o **Superficial layer; desiccated, anucleated cells; squames;** thin in **Thin skin**
- **Stratum lucidum**
  - o **Lost nuclei; poorly stained keratinocytes<u>; prominent in THICK SKIN</u>**

D. **Cells of the Epidermis**
- **Keratinocytes-**originates in the basal layer, **predominant cell type, produce keratin**
- **Melanocyte-** produce and **distribute brown to black melanin into keratinocytes**
- **Langerhans' cells- dendritic antigen presenting cells**
- **Merkel's cells-sensitive mechanoreceptor (touch)**

E. **Dermis**
- Predominantly **dense irregular collagenous connective tissue**
  - o <u>**Three Layers**</u>
    - ▪ **Papillary-loosely woven, capillary loops, Meissner's Corpuscles (encapsulated nerve endings)**
    - ▪ **Reticular-coarser; vascular**; collagen fibers, hair follicles, sebaceous and sweat glands
    - ▪ **Hypodermis-loose CT;** may be replaced with **adipose tissue (panniculous adiposus)**

F. Derivatives of Skin
- Hair Follicle and Hair
  - o Hair Follicle
    - Innervated by **sensory nerve fibers**
    - **Base** of hair follicle **forms hair root; hair root penetrated by dermal papillae**
    - **Root** and **papillae** form **hair bulb**
    - Follicle surrounded by **(glassy) hyaline** membrane
  - o Hair
    - Made of **Keratinized cells**
    - Base of hair follicle called **Hair matrix: growth**
    - Hair above skin's surface called **Shaft**
    - <u>**Two Types**</u>
      - * **Vellus**
      - * **Terminal**
  - o Cross Section of Hair and Follicle (Interior to Exterior)
    - **Medulla-**
    - **Cortex-**
    - **Cuticle-**
    - **Internal root sheath-**
    - **External root sheath-**
    - **Glassy membrane-**
    - **Connective Tissue Sheath-**
- **Sebaceous glands-**Secrete oily sebum
- **Arrector pili muscle- Smooth muscle attached to hair follicle; contraction**
- **Sweat glands**
  - o **Simple coiled tubular glands**
    - **Eccrine:** watery secretion
    - **Apocrine:** viscous secretion
  - o **Myoepithelial cells**

G. Nails
- **Keratinized (Cornified) structures**
  - o **Nail plate-** main part
    - Distal free edge
    - **Nail root**
- **Lie on Nail bed-**epidermal layer
  - o **Nail matrix-**produces the nail
- **Eponychium-cuticle**
- **Hyponychium-beneath the free edge of nail bed**
- **Lunula- opaque, crescent shaped area of nail plate**

H. Special Receptors
- <u>**Epidermis**</u>
  - o **Merkel's disc: Nonencapsulated; Mechanoreceptors**
  - o **Krause's bulbs: Encapsulate;** deep to the epidermis; **Function not known**
  - o **Thermoreceptors: Nonencapsulated:** Respond to **Temperature**
  - o **Nociceptors: Nonencapsulated;** Respond to **Pain**
- <u>**Dermis**</u>
  - o **Meissner's corpuscles: Encapsulated; dermal papillae;** Respond to **Touch Sensations**
  - o **Ruffini's endings: Encapsulated;** nail beds; Respond to **Pressure and Touch**
  - o **Pancinian corpuscles: Encapsulated; hypodermis;** Respond to **Vibration, Pressure, Deep Touch**

HISTOLOGY

RESPIRATORY SYSTEM

A. **Respiratory System**
   - Consist of <u>Conducting Airways</u> and <u>Respiratory Airways</u>

B. **Conducting Airways**
   - **Adjust Temperature, Humidify, Clean the air**
     - o Consists of:
       - ▪ **Nasal passages**- mostly **pseudostratified ciliated columnar (PSCC) epithelium with goblet cells** and **CT sheath** with **seromucous glands**
       - ▪ **Olfactory epithelium**: thicker, **lacks mucous glands; bipolar neurons; Bowman's glands**
       - ▪ **Larynx**: **cartilaginous framework**; lined by **PSCC** and a **propria**
         - * **Epiglottis**- **elastic cartilage** covered by **Stratified squamous epithelium (SSE)**
       - ▪ **Trachea**
       - ▪ **Paired main bronchi**

C. **Trachea**
   - **Mucosa**
     - o **Respiratory Epithelium**
       - ▪ Tall **PSCC**
       - ▪ **Mucous (goblet) cells**-produce **mucinogen→mucin**
       - ▪ **Basal cell**-regenerative cells
       - ▪ **Brush cells**: microvilli; **neurosensory functions**
       - ▪ **Serous cells**: serous secretion; function not understood
       - ▪ **DNES (Diffuse Neuroendocrine System) cells**: **hormones**; regulate respiratory function
     - o **Lamina propria**
       - ▪ **Collagen** and **elastic fibers;** lymphatic tissue, fibroblasts, macrophages
   - **Submucosa**
     - ▪ **Dense fibrous CT**
     - ▪ **Seromucous glands**
     - • **Tracheal Cartilage**
     - ▪ Uniquely **"C- shaped"**
     - ▪ **Hyaline cartilage (15-20) rings** with a **smooth muscle** bridge (**Trachealis muscle**)
     - ▪ Separates the submucosa from the adventitia
   - **Adventitia**
     - ▪ **Loose CT**

D. **Bronchi**
   - Trachea divides into two **Primary bronchi→lung→Secondary (lobar) bronchi Lobar bronchi** further divide→**Tertiary (segmental) bronchi**
   - **Extra pulmonary Bronchi**
     - o **Histology resembles Trachea**
   - **Intrapulmonary (Secondary) Bronchi**
     - o **Tall PSCC→shorter, flatter**
     - o **Cartilage <u>"rings" replaced</u> by hyaline cartilage <u>"plates"</u>**

o **Decrease in number of glands and goblet cells**
o **Increase in smooth muscle and elastic tissue**

E. **Bronchioles**
  - **Tertiary (segmental) bronchi→bronchioles**
  - **<u>Cartilage PLATES and GLANDS ABSENT</u>**
  - **Mucosa**
    o **Respiratory Epithelium**
      ▪ **Progress from Simple ciliated columnar → Simple ciliated cuboidal epithelium**
      ▪ **Goblet cells gradually disappear**
      ▪ <u>**Clara cells**</u> :dome shaped; secrets **"surface active agent"** that prevents luminal adhesion with airway collapse; **lysosomes**
    o **Lamina propria**
      ▪ **Elastic fibers and smooth muscle present**
  - **Terminal Bronchioles**
    o **Conduction part ends**
    o **Epithelium: Simple cuboidal (with some ciliated) and Clara cells**
    o **Reduced CT and smooth muscle**

F. **Respiratory Airways**
  - **Terminal bronchioles→<u>Respiratory Bronchioles</u>**
  - **Gas exchange occurs**
  - **Respiratory Bronchioles**
    o **Epithelium**: **cuboidal and Clara cells**
    o Gradually **Clara cell**s predominate
    o Possess **alveoli** (structure permitting gas exchange)
  - **Alveolar Ducts**
    o Arise **from respiratory bronchioles**
    o Elongated airways with **little or no walls**; only alveoli
  - **Alveolar Sacs**
    o Arise from ducts; are **spaces surrounded by alveoli**

4. **Alveoli**
  - Separated from each other by **Septa (contain connective tissue and capillaries)**
  - Septa is the **site of the air-blood barrier**
  - **Pores of Kohn** allow **alveolar-to-alveolar air gas circulation**
  - **Alveolar epithelium** is composed of **specialized cells**
    o **Type I Alveolar Cells (Type I pneumocytes)**
      ▪ **Extremely thin squamous cells**
      ▪ **95%** of the alveolar surface
      ▪ **Form an air- tight barrier**
    o **Type II Alveolar Cells (Septal cells, Type II pneumocyte)**
      ▪ **Can differentiate** into **Type I cells**
      ▪ **Cuboidal, secretory, bulging**
      ▪ **Makes Surfactant**
    o **Brush cells - receptor cells, microvilli**
    o **Dust cells- macrophages**

HISTOLOGY

DIGESTIVE SYSTEM I: Oral Cavity and Associated Structures

A. Digestive system
- Consists of the **teeth, tongue, salivary glands, esophagus, stomach, small and large intestines, pancreas, gallbladder, liver**
- Functions in digestion, absorption, elimination of food and unusable parts

B. Oral Cavity
- Consists of the lips, mouth, teeth, tongue, salivary glands
  - o **Lips**
    - **External**-Epidermis and dermis (hair follicles, sebaceous and sweat glands)
    - **Transitional (Vermillion) zone**-pink area; absent hair and sweat glands
    - **Internal-Stratified, squamous, nonkeratinized epithelium**
    - **Core-Skeletal muscle** and fibroelastic **CT**
  - o **Oral Mucosa (Mucous Membrane)**
    - **Epithelium-Stratified Squamous nonkeratinized epithelium**
    - **Lamina propria**-loose CT
    - **Submucosa**-denser CT
  - o **Teeth**
    - 20 **deciduous** replaced by 32 **permanent**
    - Consist of a **crown, root, pulp cavity**
    - <u>**Three Calcified Tissues**</u>
      - * **Enamel: Hardest substance in body; 96% calcium hydroxyapatite**; produced by **ameloblasts**
      - * **Dentin**: Internal; **70% calcium hydroxyapatite** and **collagen fibers;** produced by **odontoblasts**
      - * **Cementum: 45% calcium hydroxyapatite** and **collagen fibers**; covers **Root** of tooth; produced by **cementoblasts**
    - **Pulp Cavity: gelatinous** ; mesenchymal connective tissue; **Nerves and Blood vessels**
    - **Periodontal Ligament**-Dense CT attaching connects **cementum** with alveolar bone
  - o **Gingiva**
    - **Masticatory mucosa- Stratified squamous <u>parakeratotic</u> epithelium.**
  - o **Tongue**
    - **Mucosa membrane-Stratified squamous epithelium** and **lamina propria**
    - **Skeletal muscle**
    - **Four types of lingual papillae(modified mucosa)**
      - * **Filiform: Most numerous; <u>lack taste buds</u>;** long slender
      - * **Fungiform: Taste buds;** mushroom shaped
      - * **Foliate: Taste buds**; side of tongue
      - * **Circumvillate-Taste buds;** V shaped
    - **Taste Buds: Neuroepithelial cells**; recognizes: **bitter, salty, sour, sweet, or umani**
  - o **Major salivary glands**
    - Produce **saliva**: Protection, digestion; **water, mucus, enzymes, electrolytes, antibodies**
    - <u>**Three Main Major Salivary Glands**</u>
      - * **Parotid: Largest; Serous only**
      - * **Submandibular-Mixed; mostly serous**
      - * **Sublingual-Mixed; mostly mucous**

HISTOLOGY

DIGESTIVE SYSTEM II: Alimentary Canal

A. **Digestive System**
- **Alimentary canal** travels from the **oral cavity to the anus; Performs task in digestion**
- **Hollow Tube made of Four Layers**
  - **Mucosa**
    - Inner most layers. Made up of :
      - * **Epithelial lining**-protection, secretion, absorption
      - * **Lamina propria**-connective tissue, glands, vessels
      - * **Muscularis mucosae** –usually **smooth muscle** layers; motility
    - Protection barrier
    - **Absorption**-movement of nutrients, water, electrolytes into blood and lymphatic vessels
    - **Secretion**-lubricates and delivers digestive enzymes, hormones, antibodies into the lumen
  - **Submucosa**
    - **Dense irregular connective tissue**
    - Blood vessels; lymphatic vessels; **nerve plexus (of Meissner)**; glands
  - **Muscularis externa**
    - **Two layers of smooth muscle**; Contract: mix and propel contents
      - * **Inner circular layer**
      - * **Outer longitudinal layer**
      - * **Myenteric (orAuerbach) nerve plexus**
  - **Serosa or Adventitia**
    - Outer most layer
    - **Adventia-CT** that attaches portions of the digestive tract to the abdominal and pelvic cavities
    - **Serosa**-Thin layer **simple squamous epithelium** called **mesothelium**

B. **Esophagus**
- Fixed muscular tube which delivers food and liquid to the stomach
  - **Mucosa**
    - **Epithelium-Nonkeratinized Stratified Squamous Epithelium**
    - **Lamina propria:** diffuse lymphatic tissue; **mucus-secreting esophageal-cardiac** glands
    - **Muscularis mucosa: smooth muscle**
  - **Submucosa**
    - Dense collagenous CT mixed with elastic fibers; **esophageal mucus-secreting glands; nerve plexus**
  - **Muscularis externa**
    - **Inner circular layer**
    - **Outer longitudinal layer**
    - Upper 1/3- Skeletal muscle
    - Middle 1/3- Skeletal and smooth muscle
    - Distal 1/3- Smooth
    - **Nerve plexus**
  - **Adventitia**

- **Esophageal-gastric Junction**
  - o **Mucosa**
    - **Nonkeratinized Stratified Squamous Epithelium→Simple Columnar Epithelium**
    - **Lamina propria**: occupied by **gastric (cardiac glands)**

C. Stomach

- Receives macerated food from the esophagus; mixes and partially digest food producing **chime**
- Divided into four regions :**cardiac, fundus, body** (corpus)**, pyloris**
  - o **Mucosa**
    - **Epithelium-Simple Columnar Epithelium;** composed of **surface mucous cells** with flattened basal nucleus**; apical cytoplasm filled with alkaline mucus**
      Depressions called **gastric pits→**gastric glands (isthmus, neck, body)
      <ins>**Cell Types in glands**</ins>
      - * **Mucous neck cells**: cuboidal to irregular shape; **water-soluble mucus**
      - * **Chief (zymogenic) cells: blue** on **H&E**; protein secreting cell; **secrete** precursors of enzymes **pepsin, renin**; secrete **lipase**
      - * **Parietal cell:** large triangular shaped cells with spherical nucleus; **red** on **H&E**; secretes HCL **and intrinsic factor (IF)**
      - * **DES (Enteroendocrine (neuroendocrine)) cells: pale staining** on **H&E**; releases **hormones (secretin, VIP, gastrin, somatostatin…)**
      - * **Stem (regenerative) cell;** undifferentiated; replace lining and cells
    - **Lamina propria- Gastric glands**
    - **Muscularis mucosa**-Thin layers **smooth muscle**
  - o **Submucos**a-Dense connective tissue, adipose tissue, blood vessels, **nerve plexus**
  - o **Muscularis Externa**-Randomly arranged
    - **Inner obligate layer**
    - **Middle circular layer**
    - **Outer longitudinal layer**
    - **Myenteric (orAuerbach) nerve plexus**
  - o **Serosa**
    - (Pyloric region stomach)-**Adventitia**

D. Small Intestine

- **Longest component** of the digestive system
- Principle site for **digestion and absorption**
- Three Regions
  - o **Duodenum**-receives chime; enzymes
  - o **Jejunum**-principle site of absorption of nutrients
  - o **Ileum**-water and electrolyte reabsorption
- **Gross-plicae circulares** (circular folds)
- All Three regions displays surface **Villi (mucosal** extensions of **lamina propria)**
- **Villi** extend into intestinal lumen; **increase surface area for absorption**
- **Intervillous spaces between villi**
- Intestinal glands **(crypts of Lieberkühn)**: tubular glands which open into intervillar spaces
  - o **Mucosa**
    - **Epithelium: Simple Columnar Epithelium** with **striated border (microvilli)**
      <ins>**Cell Types**</ins>
      - * **Enterocytes: Tall absorptive columnar cells** with **microvilli; enzymes in apical membrane; glycocalyx**
      - * **Goblet cells**: Clear on H&E; **Mucus producing cells**

* **Paneth cells: secrete antibacterial substances (lysozymes)**; apical **red granules**
* **DES (Enteroendocrine (neuroendocrine)) cells: hormones** (CCK, secretin, Motilin, etc.)
* **M (Microfolds) cells**-Modified; assist in **immune response (endocytosis)**

D. Small Intestine (Cont.)
- o **Mucosa (Cont.)**
  - ▪ **Lamina propria**
    * Loose CT; **capillary bed, lymphatic lacteal,** lymphocytes and macrophages; houses **crypts of Lieberkühn**
    * <u>Ileum</u>-lamina propria contain large accumulations of **lymphatic nodules** called <u>**Peyer's Patches**</u>
  - ▪ **Muscularis mucosa: inner circular and outer longitudinal smooth muscle(32)**
- o **Submucosa-Fibroelastic CT**
  - * **In <u>Duodenum</u>**- numerous **alkaline, mucin –containing glands** called **Brönner's glands**
- o **Muscularis Externa: Inner circular and outer longitudinal; Auerbach's plexus in between**
- o **Adventitia and Serosa**
  - ▪ **Adventitia-** distal duodenum
  - ▪ **Serosa: 1st part of duodenum ;jejunum; ileum**

E. Large Intestine
- • **Reabsorption** of electrolytes and water and the elimination of undigested food and waste
- • **No plicae circulares or villi**
- • Divided into: **cecum, appendix, colon** (ascending, transverse, descending, sigmoid), **rectum, anal canal**
  - o **Mucosa**
    - ▪ **Epithelium: Simple Columnar Epithelium with goblet cells**
      <u>**Cell Types**</u>
      * **Enterocytes**: Reabsorption cells; **tall columnar cells**; nuclei at base; **microvilli**
      * **Goblet cells: abundant;** produce **mucin**
      * **Paneth cells-**<u>absent</u>
      * **DES (Enteroendocrine (neuroendocrine)) cells: hormones** (serotonin, SubstanceP.)
      * **M (Microfolds) cells**-Modified; assist in **immune response (endocytosis)**
    - ▪ **Lamina propria-**houses **crypts of Lieberkuhn**
    - ▪ **Muscularis mucosa-prominent**
  - o **Submucosa: Fibroelastic CT;** no glands
  - o **Muscularis Externa**
    - ▪ **Inner circular layer-**unremarkable
    - ▪ **Outer longitudinal layer: modified;**
    - ▪ Three **thicken equally spaced bands called** <u>Teniae coli</u>
  - o **Adventitia and Serosa**
    - ▪ **Adventitia:** ascending and descending colon; anal canal
    - ▪ **Serosa:** cecum; appendix; transverse colon; sigmoid colon; rectum
      * Fat filled pouches **(appendices epiploicae)**

F. Vermiform Appendix
- • Arises from **cecum**
- • Lumen is **stellate shaped**
  - o **Mucosa**

- **Epithelium-Simple Columnar Epithelium**
- **Lamina propria:** abundant **lymphatic nodules; some crypts of Lieberkühn**
- **Muscularis mucosae-**smooth muscle
  - o **Submucosa: Fibroelastic CT; lymphocytes**
  - o **Muscularis Externa**
    - **Inner circular and Outer longitudinal layers**
  - o **Serosa**

G. **Rectum and Anal Canal**
- **Rectum**
  - o Distal part of the **Large Intestine**
  - o **Continuation** of the **sigmoid**
  - o **Ends at the anal canal**
  - o **Mucosa**
    - Resembles rest of large intestine
    - **Crypts shallower; more goblet cells**
- **Rectoanal Junction**
  - o **Mucosa**
    - **Epithelium: Simple Columnar Epithelium →Stratified Squamous Epithelium**
- **Anal Canal**
  - o **Terminal portion** of the large intestine
  - o Presents **anal columns** which join at the orifice of the **anus** to form **anal valves** and **anal sinuses**
  - o **Mucosa**
    - **Epithelium: Stratified Squamous Epithelium<u>Epidermis</u> (at the orifice)**
  - o **Muscularis externa**
    - Forms the **internal anal sphincter muscle**
  - o **Adventitia**

HISTOLOGY

DIGESTIVE SYSTEM III: Major Glands

A. **Digestive system**
  - Located outside the walls of the alimentary track
  - **Connected** to lumen of the track **via ducts**

B. **Major salivary glands (Discussed)**
  - **Secrete saliva (watery composition of mucus, enzymes, ions, antibodies)**
    o **Parotid**
    o **Submandibular**
    o **Sublingual**

C. **Liver**
  - **Largest internal organ**
  - Produces most of the **body's plasma proteins**
  - Stores **vitamins A,D,K**
  - **Detoxification** center for degrading alcohol and drugs
  - Stores **glycogen**
  - **Secretes bile salts, glucose**
  - **Duel** blood supply
    o **Hepatic portal vein**-receives **venous(nutrient rich) blood** from **GI system**
    o **Hepatic artery**-receives **arterial (oxygenated) blood** from **celiac trunk**
    o **Both vessels enter via the Porta Hepatis**
  - <u>**Histology**</u>
    o **Glisson's Capsule**
      ▪ **Invests liver;** sends **septa** to subdivide **parenchyma** into **lobules**
    o **Parenchyma**
      ▪ **Hepatocytes: stacked;** arranged in **cords; cuboidal** in shape; **red cytoplasm (H&E)** due to **glycogen**
      ▪ One cell thick; separated by **hepatic sinusoids (sinusoidal capillaries)**
    o **Connective Tissue Stroma**
      ▪ Blood vessels, nerves, lymphatic vessels, bile ducts
    o **Sinusoidal Capillaries**
      ▪ **Vascular channels** between the plates of the hepatocytes
      ▪ Lined with **endothelial cells and**
      ▪ **Kupffer cells (Macrophages)**
    o **Perisinusoidal Spaces (Spaces of Disse)**
      ▪ **Between** the **sinusoidal endothelium** and the **hepatocytes**
      ▪ **Ito (Fat-storing) Cells**-accumulate and store **Vitamin A;** produce **Type I collagen**
      ▪ Site of **exchange of materials** between blood and liver cells
    o **Bile Canaliculi**
      ▪ Intercellular channels carry **bile produced by hepatocytes**
      ▪ **Bile** empties into **bile ducts**
      ▪ **Bile ducts** converge to **hepatic ducts**

- **Histology (Cont.)**
  - o **Central Vein**
    - ▪ **Where mixed arterial and venous blood drains**; this blood leaves liver through **hepatic veins**→empty **into IVC**
  - o **Classic Liver (Hepatic) Lobule**
    - ▪ Anastomosing plates of hepatocytes arranged to form **hexagonal lobules**
    - ▪ Where **lobules** meet, their **CT merge to form portal areas** which house **Portal Triad**
      - * **Hepatic artery**
      - * **Portal vein**
      - * **Bile duct**
      - * Lymphatic vessels
      - * Branches of **Vagus nerve**
    - ▪ **Center** of lobule houses a **central vein**
  - o **Portal Lobule**
    - ▪ **Center** of liver lobule is **bile duct**
  - o **Acinar Lobule**
    - ▪ Blood flow into **adjacent classic lobules**
    - ▪ **Center** is **shared side** with **portal triads** at each end

D. **Gallbladder**
  - • **Stores and concentrates bile**; empties into duodenum; pear shaped
  - • Connected to liver via cystic duct, which joins the **common hepatic duct**
  - • **Histology**
    - o **Mucosa**
      - ▪ **Epithelium-Simple Columnar Epithelium with microvilli**
      - ▪ **Lamina propria: Rokitansky-Aschoff sinuses;** lymphocytes, plasma cells
      - ▪ **Muscularis mucosae-absent**
    - o **Submucosa-absent**
    - o **Muscularis externa**
      - ▪ Irregularly arranged **smooth muscle fibers**
    - o **Adventitia and Serosa**
      - ▪ **Adventitia**-attaches gallbladder to capsule of liver
      - ▪ **Serosa**-covers remaining surface

E. **Pancreas**
  - • Will discuss in Endocrine
  - • Empties into **duodenum** via **pancreatic duct**

HISTOLOGY

URINARY SYSTEM

A. **Urinary system**
  - Consists of: **two kidneys, two ureters, bladder, urethra**
  - **Produces, stores, voids urine**
  - **Regulates blood pressure and fluid volume**
  - Makes and releases **certain hormones**

B. **Kidney**
  - Eliminates body's waste products
  - Regulates body's ion balance and water content
  - Role in stabilizing **Blood Pressure (renin)**
  - Make **erythropoietin (hormonal regulator of RBC production)**
  - Hydroxylation of **25-OH Vitamin D3**

C. **Structure of kidney**
  - **Capsule**-Two layers of **dense fibrous CT**
    - o **Outer:** paler; **fibroblasts;** blood vessels
    - o **Inner:** darker; thinner; **myofibroblasts;** blood vessels
    - o Surrounded by **perirenal fat**
  - **Renal Cortex (Outer Portion)**
    - o **Cortical Labyrinth** consist of:
      - ▪ **Renal corpuscles** (layers of **Bowman's capsule, Bowman's space, glomerulus, mesangial cells**)
      - ▪ Cross sections of **proximal and distal convoluted tubules; macula densa** section of **distal tubules**
    - o **Cortical Medullary rays**
      - ▪ Extension of the **renal medulla** into the cortex
      - ▪ Mostly **collecting tubules; pars recta of proximal tubules; ascending thick limbs of Henle's loop;** blood vessels
  - **Renal Medulla (Inner Portion)**
    - o Composed of **Pyramids**
      - ▪ Consist of **collecting tubules; vasa recta;** renal interstitium
  - **Hilum**-renal **vessels and nerves** pass through
  - **Pelvis:** divided **into minor and major calyxes; joins the ureter**
  - **Vascular supply:** Abdominal aorta→Renal artery(R and L)→Interlobar arteries→Arcuate arteries→Interlobular arteries→Afferent glomerular arterioles→Glomerulus→Efferent glomerular arterioles→Interlobular veins→Arcuate veins→Interlobar veins→Renal veins→IVC

D. **Functional Unit of the Kidney**
  - **Uriniferous Tubule**
    - o Consists of <u>**Nephron**</u> and **Collecting Tubules**
      - ▪ <u>**Nephron**</u>
        - * Classified by location of renal corpuscles
          -**Juxtamedullary, Cortical (subcapsular), Midcortical (intermediate)**
        - * Responsible **for filtration, excretion, resorption**

D. **Functional Unit of the Kidney (Cont.)**
- **Uriniferous Tubule**
  - o Consists of <u>**Nephron**</u> and **Collecting Tubules**
    - ▪ <u>**Nephron**</u>
      - * <u>**Renal Corpuscle**</u>
        1. **Bowman's Capsule**
           - **Parietal layer** (outer layer)-**Simple Squamous Epithelium**
           - **Visceral layer**(inner layer)-**Podocytes** with processes called **pedicels** which wrap around glomerular capillaries
        1. **Bowman's Space**
        2. **Glomerular Capillaries**
           - **Fenestrated** with large pores
           - Three layered **basal lamina**
           - * Lamina rara externa; Lamina densa Lamina rara interna
        3. **Mesangial Cells**
           - Intraglomerular-**phagocytic; contractile**
           - Extraglomerular- **phagocytic; contractile**
      - * <u>**Proximal Tubule**</u>
        1. **Proximal <u>Convoluted</u> Tubule (PCT) segment**
           - **Simple Cuboidal Epithelium**; brush border (**dense microvilli**), star shaped lumen
        2. <u>**Proximal Straight Tubule**</u> **(pars recta of the proximal tubule) segment**
           - **Simple Cuboidal Epithelium**; brush border (**microvilli**)
           - Also referred to as the **descending <u>thick</u> limb of Henle's Loop**
      - * <u>**Loop of Henle (LOH)**</u>
        1. Composed of **Simple Squamous Epithelium**
        2. **Three Regions**
           - **Descending <u>thin</u> limb** (continuation of **proximal straight tubule**)
           - **Henle's loop-hairpin; (**connects descending and ascending limbs)
           - **Ascending <u>thin</u> limp** (becomes **distal tubule**)
      - * <u>**Distal Tubule**</u>
        1. **Ascending <u>thick</u> limp of Henle's Loop (pars recta of the distal tubule)**
           - **Simple Cuboidal Epithelium <u>lacking</u> a brush border**
        2. **Distal <u>Convoluted</u> Tubule**
           - **Simple Cuboidal Epithelium <u>lacking</u> a brush border**
           - Shorter than PCT
        3. **Macula Densa**
        - **Modified** cells in contact with the **afferent and efferent glomerular arterioles**
        - Thin, tall **cuboidal** cells
        - Communicates with **juxtaglomerular (JG) cells** (modified smooth muscle cells)
        4. **Juxtaglomerular (JG) apparatus**
           - Consist of the **macula densa, JG cells, extraglomerular mesangial cells**
           - Maintains **BP**

D. **Functional Unit of the Kidney (Cont.)**
- **Uriniferous Tubule**
  - Consists of **Nephron** and <u>**Collecting Tubules**</u>
    - <u>**Collecting Tubules**</u>
      - * Begin at terminal ends of **DCT**
      - * Consists of **Simple Cuboidal Epithelium**
      - * <u>**Two Cell Types**</u>
        1. **Principle cells (light):** Single, **nonmotile cilium; ADH-sensitive channels**
        2. **Intercalated cells** (dark)
           - Type A cells (secrete H+) and
           - Type B cells (resorb H+ and secrete HCO3-
      - * Empties into a <u>**Collecting Duct**</u>
    - <u>**Collecting Ducts**</u>
      - * **Cortical Collecting Ducts** joins to **form Medullary Collecting Ducts**
      - * **Both of Simple Cuboidal Epithelium**
      - * Form **papillary ducts (of Bellini)-Simple Columnar Epithelium**
      - * <u>**Two cell types**</u>

E. **Renal Circulation**
- Aorta→blood enters kidney through renal artery→ segmental arteries→lobar arteries→interlobar arteries→arcuate arteries→interlobular arteries→afferent arterioles→ glomerul (capillaries)→drain by efferent arterioles→peritubular capillaries and vasa recta→interlobular veins→arcuate veins→interlobular veins→renal vein→IVC

F. **Calyces and Renal Pelvis**
- Papilla projects into a minor calyces→form major calyces→empty into renal pelvis
- All lined by **Transitional Epithelium** and lamina propria; **smooth muscle** deep to lamina propria

H. **Ureters**
- Composed of **Three Layers**
  - **Mucosa: Epithelium-**<u>**Transitional Epithelium**</u>**; Lamina propria-fibroelastic CT**
  - **Muscularis: Two smooth muscle layers (Inner-longitudunal, Outer-circular)**
  - **Adventitia**

G. **Bladder**
- Receives urine from the ureters and stores until time to void (urinate)
- Composed of **Three Layers**
  - **Mucosa: Epithelium-**<u>**Transitional Epithelium**</u>**; Lamina propria-fibroelastic CT**
  - **Muscularis: Three smooth muscle layers (Inner-longitudunal, Middle-circular, Outer-longitudinal); detrusor muscle**
  - **Adventitia and Serosa**

H. Urethra
- Carries urine from bladder to external environment
- **Fibromuscular tube**
- **Female**
    - **Mucosa: Epithelium**-mostly **Nonkeratinized Stratified Squamous Epithelium** with **Transitional Epithelium** (proximal end) and patches of **Pseudostratified Columnar Epithelium; Lamina propria-** paraurethral **glands of Littre**
    - **Muscularis: Smooth muscle (Inner longitudinal** and **Outer circular)**
    - **Muscularis** surrounded by **skeletal muscle** at external urinary sphincter
- **Male**
    - **Three Segments**
        - **Prostatic urethra: Transitional Epithelium**
        - **Membranous urethra: Stratified Columnar Epithelium** with regions of **Pseudostratified Epithelium**
        - **Spongy (penile urethra): Stratified Squamous or Pseudostratified Columnar Epithelium** with **Stratified Squamous** at the orifice; **Lamina propria-glands of Littre**

HISTOLOGY

ENDOCRINE SYSTEM

A. **Endocrine System**
   - Consist of endocrine **glands and cells**
   - Typically **epithelial in origin; ductless; very vascular**
   - **Produces hormones** (regulate activities of various cell, tissue, organs)
     o   Three Types of Hormones
       ▪ Nonsteroidal-based (proteins and polypeptides)
       ▪ Steroid based (cholesterol derivatives)
       ▪ Amino acid derivatives
   - Helps in **maintaining homeostasis and coordinating body growth and development**
   - Bind to hormone receptors on its "target cell"
   - **Hypothalamus coordinates most endocrine functions of the body**

B. **Hypothalamic-Pituitary System**
   - **Hypothalamic Gland**
     o   Located at **base of brain**
     o   Connected to the anterior pituitary by portal hypophyseal blood vessels
     o   Connected to the posterior pituitary by way of a nerve tract Hypothalamohypophysial tract)
   - **Pituitary Gland (AKA Hypophysis)**
     o   Located in the sella turcica
     o   **Divided into anterior and posterior lobes**

C. **Hypothalamic Hormones**
   - Somatostatin **(GHIH)**
   - Corticotropin-releasing hormone **(CRH)**
   - Thyrotropin-releasing hormone **(TRH)**
   - Growth hormone-releasing hormone **(GHRH)**
   - Gonadotropin-releasing hormone **(GnRH)**
   - Prolactin-inhibiting hormone **(PIH)** and Dopamine
   - Prolactin-releasing hormone **(PRH)**
   - Antidiuretic hormone **(ADH)**
   - Oxytocin

D. **Pituitary Gland**
   - Composed **of glandular epithelial tissue and neural tissue**
     o   Growth Hormone **(GH)** aka Somatotropin **(STH)**
     o   Thyroid-stimulating hormone**(TSH)**
     o   Follicle-stimulating hormone**(FSH)** and Luteinizing hormone**(LH)**
     o   Adrenocorticotropic hormone**(ACTH)**
     o   Prolactin
     o   Melanocyte-stimulating hormone**(MSH)**

D. **Pituitary Gland (Cont.)**
- **Anterior lobe(Adenohypophysis)-Glandular Epithelium**
  - o Three Parts: **pars distalis, pars internedia, pars tuberalis**
  - o **Cell Types**
    - ▪ **Chromophils**
      - * **Acidophils (pink with H&E); GH, prolactin, STH**
      - * **Basophils (blue with H&E); ACTH, TSH, FSH, LH, ICSH**
    - ▪ **Chromophobes**
      - * **Clear (with H&E)**
      - * **?** Chromophils which released their granules
- **Posterior lobe(Neurohypophysis)-extension of the CNS-Neural Tissue**
  - o **Neurosecretory cells store and release secretory products** from the hypothalamus
  - o **Cells: Pituicytes; Herring body**
  - o **ADH (vasopressin)-controls blood pressure** (Works on **collecting tubules** in the **kidney**)

E. **Pineal Gland**
- Outgrowth from roof of the third ventricle; covered by **pia mater**
- **Cells**
  - o **Pinealocytes- Modified neurons** which produce **melatonin**
  - o **Glial cells**-physical and nutritional support to **pinealocytes**
  - o **Pineal sand (corpora arenaces-** made of calcium phosphate and calcium carbonate
- Regulates **circadian rhythm**

F. **Thyroid Gland**
- Function is **essential to metabolic rate, normal growth and development**
- **Connective Tissue Capsule** which the **septa** divide the gland into lobules
  - o **Thyroid Follicles**
    - ▪ **Simple Cuboidal or Low Columnar Epithelial Cells**
      - * **Follicular Cells**
        - - Secrete **Thyroid hormones (T3, T4)**
        - - **Lightly basophilic cytoplasm; round nuclei**
      - * **Parafollicular cells (C Cells)**
        - - Lie **within** the **follicular basal lamina**
        - - **Pale staining** cells secrete calcitonin
        - - Secrete **calcitonin**
    - ▪ **Colloid**
        - - **Gel like mass** inside the follicle
        - - Composed of **thyroglobulin (inactive storage form of T3 and T4)**

G. **Parathyroid Glands**
- **Four** located **posterior to thyroid gland**
- Surrounded **by a fibrous capsule** which sends **septa** and divide it into lobes
- Regulate serum **calcium and phosphate levels**
  - o **Epithelial Cells**
    - ▪ **Principle (Chief) cells**
        - - Secrete **PTH (regulates Calcium and phosphate)**
        - - **Small polygonal cells**
        - - Central nucleus with **pale** eosinophilic cytoplasm

- **Oxyphil cells**
  - Round with **acidophilic (red) cytoplasm**
  - Function unknown

H. **Adrenal (Suprarenal) Glands**
- Located **superior to the kidneys**
- Consist of a **fibrous CT capsule** surrounding the **parenchyma**
- Rich **vascular supply**
- Secretes both steroid hormones and catecholamines
  - **Cortex (Outer Region)**
    - **Three Zones**
      - * **Zona glomerulosa-outer zone**
        - Beneath capsule
        - Arranged in **closely packed ovoid clusters** with **dense staining nuclei**
        - Cells are **small and columnar or pyramidal** in shape
        - **Secretes mineralocorticoids (aldosterone)**
        - Modulates water and electrolyte balance
      - * **Zona fascicularis-middle zone**
        - **Long straight cords, one or two cells thick**
        - Cells called **spongiocytes**
        - **Light staining round nucleus**
        - Functions in **glucose and fatty acid metabolism**
        - **Secretes glucocorticoids (cortisol)**
      - * **Zona reticularis-inner zone**
        - **Anastomosing cords of cells** with **large pigment granules**
        - Rich **capillary network**
        - **Secretes androgens (sex hormones)** and glucocorticoids
  - **Medulla (Middle Region)**
    - **Derived from neural origin**
      - * **Chromaffin cell** arranged in short cords surrounded by **capillary networks**
      - * Cells stain **intensely dark and** consist of **granules**
      - * Produce **catecholamines (either norepinephrine or epinephrine)**
      - * Increase heart rate, blood pressure, sweating

I. **Pancreas**
- An **exocrine and endocrine** gland
  - **Exocrine pancreas**
    - **Acinar cells** produce **enzymes**
    - **Centroacinar cells** and **cells of the intercalated ducts** produce **alkaline fluid**
    - The enzymes, in an alkaline fluid, enter into the duodenum (essential for digestion)
    - This release controlled by **hormones CCK** and **secretin**
  - **Endocrine pancreas**
    - Consist of **scattered spherical aggregates of cells richly vascularized called**
    - **Islets of Langerhans**
      - * Clusters of **pale staining cells surrounded by intensely staining acini**
        - α **(alpha cells):** located **peripherally**; secrete **glucagon**
        - β **(beta cells): centrally** located; secrete **insulin**
        - δ **(delta cells):** located peripherally; secrete **somatostatin**
    - Regulate **glucose, lipid, and protein metabolism**

HISTOLOGY

MALE REPRODUCTIVE SYSTEM

A. **Male Reproductive System**
- Primary function is **to produce sperm (spermatogenesis)** and **synthesize androgens (sex hormones)**

B. **Testes**
- Paired organs which lie within the scrotum
  - o Covered by double layer of peritoneum called **tunica vaginalis**
  - o **Tunica albuginea**-Tough dense connective tissue capsule deep to **tunica vaginalis**
  - o **Tunica vasculosa**-Vascular loose CT
  - o **Lobules of the testes-** formed by septa of the tunica **albuginea**
    - ▪ <u>Seminiferous tubules</u>-Sperm production; convoluted in shape
      - * Thin CT covering called **tunica propria** containing **myoid cells** and fibroblasts
      - * <u>**Leydig's cells (interstitial cells)-**</u> located in vascular CT surrounding seminiferous tubules
        - - **Polygonal** in shape; **eosinophilic** cytoplasm; **lipids**
        - - **Reinke's crystal** (rod shaped crystals, function unknown)
        - - Secrete **testosterone**
      - * Thick <u>**Seminiferous (germinal) Epithelium**</u>
        - - **Spermatogenic cells**-differentiate into mature sperm
        - - **Spermatogonial cells** (Type A dark, Type A pale, Type B)
      - * <u>Sertoli cells</u>-**Tall Columnar cells;** pale nuclei and dense nucleoli; blood-testis barrier; **supporting (nourishing) cells;** androgen binding protein (ABP); inhibin
  - o **Straight tubules (tubuli recti)**
    - ▪ **Simple Cuboidal Epithelium; Sertoli** cells only
  - o **Rete testis**
    - ▪ **Simple Cuboidal** or **low Columnar Epithelium** to **Simple Squamous Epithelium**
    - ▪ **Single apical cilium**, few short apical **microvilli**
  - o **Efferent ductules (ductuli efferentes)**
    - ▪ **Simple Ciliated Columnar Epithelium** and **Simple Nonciliated Cuboidal Epithelium**
  - o **Ductus epididymis**
    - ▪ Coiled tube; consists of a head, body, tail
    - ▪ **Pseudostratified Epithelium with stereocilia**
      - * Composed of tall **Principle cells** (stereocilia) and short **Basal cells** (stem cells)
    - ▪ Circular **smooth muscle coat**
    - ▪ Function is storage and maturation of sperm cells; sperm acquires mobility
    - ▪ Halo cells-intra-epithelial lymphocytes
  - o **Ductus deferens (vas deferens)**
    - ▪ **Mucosa**
      - * **Epithelium: Pseudostratified Columnar Epithelium;** occasional stereocilia
      - * **Lamina propria:** thin, elastic
      - * **Muscularis: smooth muscle:** inner longitudinal, middle circular, outer longitudinal
      - * **Adventitia**
    - ▪ Continuation known as **ejaculatory duct**; enters prostate gland; empties into urethra

C. Accessory Glands
- **Seminal Vesicles**
    - o Produce and secretes a yellow viscous material containing fructose, aa's, ascorbic acid, prostaglandins
    - o **Mucosa**
        - **Epithelium: Simple or Pseudostratified Columnar Epithelium** with occasional **non ciliated Columnar cells** and short **basal cells**
        - **Lamina propria**-thin fibroelastic CT
        - **Muscularis: smooth muscle:** inner longitudinal and outer longitudinal layers
        - **Adventitia**
- **Prostate Gland**
    - o Largest accessory sex gland of the male
    - o **Vascular** dense irregular CT capsule containing **smooth muscle cells**
    - o Glands made of **Simple Columnar Epithelium to Pseudotratified Columnar with basal cells**
        - **Three Regions:** mucosal, submucosal, external (main)
    - o Secretes whitish thin alkaline fluid containing PAP, fibrinolysin, citric acid, PSA
    - o **Corpora amylacea**-prostatic concretions **(calcified prostatic secretions)**
- **Bulbourethral glands (Cowper's glands)**
    - o Secretes fluid containing galactose, galactosamine, isalic acid
    - o Thin **CT capsule;** subdivides gland into **lobules**
    - o **Simple Cuboidal** to **Columnar epithelium**

D. Penis
- Covered by **skin**, which is hairless at distal end; loose **hypodermis** present
- Three cylindrical bodies of **erectile tissue**
    - o **Corpora Cavernosa -2 dorsal masses**
    - o **Corpus spongiosum (urethrae)-1 ventral mass**
    - o **Erectile tissue** composed of **vascular spaces** lined with **vascular endothelium**
    - o Each covered by **Collagenous capsule-Tunica albuginea**

E. Semen
- **Seminal fluid and sperm**
- Involves the testes, epididymis, seminal vesicles, prostate gland, bulbourethral glands
- Semi-alkaline consisting mainly of water, proteins, sugars
- Average ejaculate is 3 ml. and normally contains up to 100 million sperm/ml

F. Sperm
- **Spermatogenic cells:** Most abundant in **germinal epithelium;** differentiate into mature sperm
- **Spermatogonial cell:** Least differentiated; located on basal lamina; Three subpopulations
    - o **Type A dark spermatogonia cells:**
    - o **Type A pale spermatogonia cells:**
    - o **Type B spermatogonia:**
- Spermatogonia→Primary spermatocytes→Secondary spermatocytes→Spermatids →Spermatozoa
- **Spermiogenesis-**Conversion to typical sperm morphology
- **Sperm cells** are composed of a :
    - o **Head: haploid nucleus; acrosome**
    - o **Neck:** short; **centrioles**
    - o **Midpiece: mitochondria (ATP)**
    - o **Tail: flagellum**

HISTOLOGY

FEMALE REPRODUCTIVE SYSTEM

A. **Female Reproductive System**
- Includes the ovaries, uterus, fallopian tubes, vagina, breasts with mammary glands
- Primary function is **to produce ova (oogenesis)** and **synthesize hormones** (estrogen and progesterone)

B. **Ovary**
- Covered by **Simple Squamous to Cuboidal Mesothelium** called **Germinal Epithelium**
- Beneath is an irregular **CT capsule** called **Tunica Albuginea**
- Divided into
  - o **Cortex:** Covered by **germinal epithelium**; contains developing <u>**Ovarian Follicles**</u>
  - o **Medulla:** Central region; **loose CT.**; **large blood** and **lymphatic vessels; hilar cells** (androgens); **nerves**

C. **Ovarian Follicles**
- Microenvironment for developing <u>**Ovarian Follicles**</u>
- Surrounded by **CT stroma**
  - o <u>**Primordial Follicle**</u>
    - Earliest stage of follicular development
    - Consist of a **primary oocyte**
    - **Single layer** of **squamous flattened follicular cells** surrounds the **primary oocyte**
  - o <u>**Primary Follicle**</u>
    - **Unilaminar Primary Follicle**
      - * **Flattened cells become single layer of cuboidal cells** surrounding the **primary oocyte**
    - **Multilaminar Primary Follicle**
      - * **Several layers** of **Follicular cells,** now called **Granulosa cells,** form **stratum granulosum**
      - * **Zona pellucida** (secreted by oocyte)-**layer between oocyte** and **adjacent follicular cells**
      - * **CT Stroma:** coalesces around follicular cells; separated by a basement membrane - **Œ** layer called <u>Theca Folliculi</u> which has Two Layers
        1. **Interna: secretory cells; secrete androgens in response to LH**
        2. **Externa: fibromuscular outer layer**
  - o <u>Secondary Follicle</u>
    - Continued proliferation of **granulosa cells**
    - **Call Exner bodies: Fluid filled spaces** between granulosa cells; **rosette-like structure**; secreted by granulosa cells; **contain hyaluronic acid and proteoglycans**
  - o <u>Mature (or Graafian) Follicle</u>
    - **Very large** well developed; ready to release **mature oocyte**
    - **Central Antrum** (filled with **liquor folliculi) contains** mature oocyte; **wall** composed of **membranous granulosa**
    - **Cumulus oophorus:** layers of **granulosa cells** surrounding the **oocyte** forming a **mount**
    - **Corona radiata: granulosa cells;** remain **surrounding the oocyte at ovulation**
    - **Theca interna** and **Theca externa** well developed
    - Extends through **the full thickness** of the **ovarian cortex**

D. **Ovulation**
- **Complex** process; associated with **LH spike**
- **Secondary oocyte** released from **Mature (or Graafian) Follicle**

E. **Corpus Hemorrhagicum, Corpus Luteum, and Corpus Albicans**
- **Corpus Hemorrhagicum**
  - Collapse of former **Mature (or Graafian) Follicle**→ fills lumen with **blood**
- **Corpus Luteum**
  - Corpus Hemorrhagicum transformed into **Corpus Luteum**
  - **Yellow glandular structure**
  - Cells Types
    - **Granulosa Lutein Cells: Large, pale staining;** secretes **progesterone, estrogens,** and **relaxin**
    - **Theca Lutein Cells: Smaller, fewer;** secretes **progesterone** and **estrogen precursor**
  - **Corpus Luteum** becomes <u>Corpus Luteum of Menstruation</u> if fertilization and implantation do not occur
- **Corpus Albicans**
  - Replaces **Corpus Luteum of Menstruation** with fibrotic, **scar tissue**

F. **Fertilization**
- Normally in the **ampulla** of the **uterine(fallopian) tube**

G. **Uterine Tubes (Fallopian Tubes)**
- Muscular Tubes leading from ovary to uterus
- Consist of Four Regions: **Infundibulum (fimbriae), ampulla, isthmus, intramural portion**
- **Layers**
  - **Mucosa**
    - **Simple Columnar Ciliated Epithelium**
    - **Lamina propria**-vascular
    - **Nonciliated (Peg cells) secretory cells**: Factors for capacitation; nutrient rich medium
  - **No Submucosa**
  - **Muscularis**
    - **Smooth muscle**
    - Inner circular and Outer longitudinal
  - **Serosa**

H. **Uterus**
- Pear shaped organ
- **Three Regions: Fundus, body, cervix**
- Consists of **Three Layers**
  - <u>**Endometrium**</u>
    - **Mucosa of the Uterus**
      - * **Epithelium: Simple Columnar Epithelium** with **glycogen-secretory glands**
      - * **Lamina propria**-vascular **stroma**
    - **Superficial** (functional) **Layer:** Supplied by **helicine (spiral/coiled) arteries;** undergoes hormone modulated cyclic changes
    - **Basal** (deep) **Layer:** supplied by **straight arteries;** next to myometrium; unchanged during cycles

H. Uterus (Cont.)
- Consists of **Three Layers**
  - o **Endometrium:** **Three Phases during Menstrual Cycle**
    - ▪ **Proliferative (follicular) phase:** **reepithelialized;** glands elongate; stroma thicker and more vascular
    - ▪ **Secretory (luteal) phase:** glands enlarge become tortuous; lumen filled with secretory products; arteries more coiled and lengthened
    - ▪ **Menstrual phase: Superficial** (functional) **layer** desquamated → **menstrual flow**
  - o **Myometrium**
    - ▪ **Smooth muscle**
      - * Inner and Outer longitudinal
      - * Middle circular
  - o **Perimetruim**
    - ▪ **Serosa (parietal peritoneum):**fundus and posterior wall majority of uterus
    - ▪ **Adventitia**: remainder of uterus

I. Cervix
- Inferior part of **uterus** and protrudes into **vagina**
- Divided into an **Endocervix** and **Ectocervix**
  - o **Endocervix-Mucus-secreting Simple Columnar Epithelium; endocervical glands**
  - o **Ectocervix-Stratified Squamous Epithelium**
- Wall of the Cervix
  - o Dense irregular **fibroelastic CT**

J. Vagina
- **Fibromuscular tube** (joins internal reproductive organs to the external environment)
  - o **Mucosa**
    - ▪ **Stratified Squamous Epithelium**
    - ▪ **Lamina propria: fibroelastic CT**; blood vessels; neutrophils and lynphocytes
    - ▪ Lacks vaginal glands
  - o **Muscularis**
    - ▪ **Smooth muscle**
      - * Inner circular
      - * Thicker Outer longitudinal
  - o **Adventitia**

K. Mammary glands (Breasts)
- Modified **Tubuloalveolar sweat glands** which lie in **subcutaneous tissue**
- Consists of numerous individual **lobes** made up of **compound tubuloalveolar glands**
- Responsible for **milk production**
  - o **Tubuloalveolar Glands**
    - ▪ Surrounded by **adipose tissue** and separated by fibrous **septa**
    - ▪ **Inactive (Resting) Glands**
      - * **15-20** irregular **lobes** separated by **dense irregular collagenous CT**
      - * Numerous **ducts**
      - * **Sparse glandular component**
      - * **Simple Columnar Epithelial Cells**
      - * **Myoepithelial cells**
      - * **Few** lymphocytes and plasma cells

K. Mammary glands (Cont.)
- o **Tubuloalveolar Glands**
  - ▪ **Active(Lactating) Glands**
    - * **Glands branch** →**alveoli with secretions (colostrum)**
    - * **Simple Cuboidal Epithelial Cells**
    - * **Decrease in** amount of **CT** and **adipose tissue**
    - * **More** lymphocytes and plasma cells
  - o **Lactiferous duct** drains **gland** to a **lactiferous sinus**→**nipple**
- • **Nipple**
  - o Thin Epidermis- **Stratified Squamous Keratinized Epithelium**
  - o **Absent** of **hair or sweat glands**
  - o **Collagenous CT** core with lactiferous ducts which are surrounded by smooth muscle
- • **Areola**
  - o **Pigmente**d region
  - o Surrounds the nipple
  - o Contains **sweat, sebaceous, and areolar glands**

L. Placenta
- • **Very vascular structure** permitting exchange of oxygen, nutrients, and waste products between maternal and fetal circulatory systems
- • Derived from **maternal (endometrium) and fetal (trophoblast) tissues**
  - o **Trophoblasts (→syncytiotrophoblasts and cytotrophoblasts)** form the **Chorion**
  - o **Endometrium** in contact with **Chorion**→modified forming **Decidua**
  - o **Decidua :Three Regions**
    - ▪ **Decidua basalis**
      - * **Endometrial layer;** richly **vascularized** maternal side
      - * Large **polygonal decidual cells**; filled with lipid and glycogen; **villi; intervillous spaces**
    - ▪ **Decidua capsularis**
      - * **Thin layer** overlying the embryo; separates lumen of uterus from embryo
    - ▪ **Decidua parietalis**
      - * Tissue between uterine lumen and myometrium

M. Umbilical Cord
- o Covered by **amniotic membrane**
- o Filled with **Wharton's jelly** (contains **mesenchymal cells**
- o Two umbilical arteries
- o One umbilical vein

HISTOLOGY

EYE and EAR

A. **Special Sensory Organs**
- Highly **specialized Sensory Receptors** in which the **'Neural' Receptors** are part of a **'Non- Neural' Structure**
- Includes the visual, audio-vestibular apparatus, gustatory, and olfactory special sense organs

B. **The Eye**
- Highly specialized organ which **converts light energy** into **nerve action potentials**

LIGHT rays→CORNEA→LENS→RETINA→OPTIC NERVE→BRAIN

- Movable by **Extrinsic Skeletal Muscles**
- Consist of **Three Layers**
  - o **Outer Layer (Corneo-sclera Layer)**
    - Forms a tough **fibroelastic capsule**
      - * **Cornea (Five Layers)**
        1. **Stratified Squamous Epithelium, Nonkeratinized**
        2. **Bowman's Membrane-acellular**, homogenous
        3. **Stroma- dense regular collagenous CT**, fibroblasts, lymphoid cell
        4. **Descemet's Membrane-**thick basal lamina
        5. **Corneal Endothelium-Simple Squamous-to-Cuboidal Epithelium**
      - * **Sclera (Three Layers)**
        1. **Episcleral Tissue- blood vessels**
        2. **Stroma-dense regular collagenous CT**
        3. **Suprachoroid lamina-loose CT, fibroblasts, melanocytes**
  - o **Middle Layer (Uveal Layer)**
    - Very vascular
      - * **Choroid membrane (Four Layers)**
        1. **Suprachoroid-** fibroblasts and **melanocytes**
        2. **Vascular-large vessels**
        3. **Choriocapillary-capillaries**
        4. **Glassy membrane (of Bruch)-basal lamina, elastic and collagen fibers**
      - * **Ciliary Body**
        1. **Aqueous humor-forming ciliary processes; smooth muscle**
      - * **Iris (Three Layers)**
        1. **Simple Squamous Epithelium**
        2. **Fibrous-melanocytes and fibroblasts**
        3. **Pigmented Epithelium**
           - **Center** forms the **pupil of the eye**
           - Separates the **anterior chamber from posterior chamber**
           - **Intrinsic Smooth Muscles (sphincter pupillae, dilatator pupillaes)** adjust aperature of the **iris**
      - * **Ciliary Smooth Muscles**
        - Change **shape of lens (accommodation);** far and near vision

B. The Eye (Cont.)
- o **Inner Layer (Retinal tunic)**
  - ▪ **Pars Optica (Ten Layers)**
    1. **Pigment epithelium**-pigmented cells
    2. **Lamina of Rods and Cones**- rod and cone processes(photoreceptor cells)
    3. **Outer limiting membrane** -eosinophilic membrane
    4. **Outer nuclear layer**- densely packed nuclei
    5. **Outer plexiform layer**- synaptic connections between axons of photoreceptor cells and processes of bipolar and horizontal cells
    6. **Inner nuclear layer**- cell bodies of Muller, amacrine, bipolar, horizontal cells
    7. **Inner plexus layer**- synapses between ganglion cells and bipolar cells
    8. **Ganglion cell layer**- cell bodies of the optic tract neurons
    9. **Optic nerve fiber layer**- unmyelinated axons of the ganglion cells which are collected as the optic nerve
    10. **Inner limiting membrane**- innermost layer; terminal processes of Muller cells
  - ▪ **Pars Ciliaris and Pars Iridica Retinae**
    1. **Thin Columnar Epithelial Layer and Pigmented Layer**
- • **The Eye Consists of:**
  - o **Lens**
    - ▪ **Biconvex, transparent, elastic structure; lens cells**
    - ▪ **Three Layers**
      1. **Capsule (basement membrane)**-elastic
      2. **Simple Cuboidal Epithelium**-anterior
      3. **Lens fibers**-modified epithelial cells
  - o **Optic Nerve (unmyelinated axons)**
  - o **Optic Disc (aka Blind Spot: fibers of the nerves exit the orbit)**
  - o **Fovea centralis**-depression in retina **(rich in cones; greatest visual acuity)**
  - o **Ciliary Body**-secretes **aqueous fluid**
  - o **Canal of Schlemm**-Endothelium
  - o **Anterior and Posterior chambers**-contain aqueous fluid
  - o **Conjunctiva**-a mucous membrane
  - o **Eyelid**-fibroelastic plate; smooth muscle
  - o **Lacrimal gland**-produces a lysozyme-rich serous fluid which has an alkaline pH

C. The Ear
- • Specialized organ functions **to receives 'sounds' and maintain equilibrium**

  EXTERNAL EAR→TYMPANIC MEMBRANE→MIDDLE EAR→INNER EAR

- • **Divided into Three Parts**
  - o **External Ear**
    - ▪ Receives sound waves funneled onto **the ear drum (tympanic membrane)**
    - ▪ **Auricle (pinna)- elastic cartilage** covered by **skin**
    - ▪ **External auditory meatus (canal)-**skin (sebaceous and apocrine sweat glands, cerumen, and bone (temporal)
    - ▪ **Tympanic membrane**-separates external ear from middle ear

C. The Ear (Cont.)

o   **Middle Ear**
  - Consists of **the tympanic cavity (filled with air)**
  - **Houses <u>Three Auditory Ossicles (Bones)</u>**
    1. **Malleus (hammer)**-outer most
    2. **Incus (anvil)**-middle
    3. **Stapes (stirrup)**-inner most
  - **All Three articulate by synovial joints**
  - **Cavity communicates interiorly with the nasopharynx by the auditory (Eustachian) tube**
  - **Sound waves are funneled by the auricle to the tympanic membrane, then to the oval window of the cochlea**

o   **Inner Ear**
  - Concerned with **hearing and balance**
  - Consists of **membranous labyrinth (fluid called endolymph)** in the temporal bone; **Simple Epithelium** and an **osseous(bony)labyrinth (filled perilymph)**
  - **Cochlea (bone)** closest to the middle ear contains the parts responsible for **hearing and balance**
  - **Bony (osseous) labyrinth** divided into **Three Areas**
    1. **Vestible** (space)- contains **the utricle and saccule** (contain **neuroepithelial cells)**

    2. **Semicircular canals**-contain **sensory receptors with axons**

    3. **Cochlea**-Contains **the auditory sense organ**
       <u>Three Compartments</u>
       a. **Scala vestilbuli (SV):** superior
       b. **Scala media (SM)/Cochlear duct-Organ of Corti**
       c. **Scala tympani (ST):** inferior